Subaru:
From Snowdrifts to Street Legends

By

Jack Mercer

This edition published July 2025.

"The road less traveled isn't just a route—it's a philosophy."

Table of Contents

Foreword

From the moment a Subaru first rolled off the production line in postwar Japan, it was clear this was a company apart. The name itself—a reference to the Pleiades star cluster—hinted at both ambition and unity, a constellation of ideas drawn together from the far-reaching experiences of engineers whose earliest work soared through the skies. Subaru's journey, stretching from its origins as part of Fuji Heavy Industries to its position as one of the world's most distinctive and beloved automakers, is one of quiet innovation, unorthodox choices, and an enduring dedication to meeting the needs of ordinary drivers in extraordinary ways.

Automotive history is often defined by giants: household names that dominate sales charts and saturate city streets. Yet, every so often, a brand emerges not by mimicking trends, but by charting a course of its own. Subaru has made a habit of thriving at the margins, favoring engineering clarity over fleeting fashion, and earning a fiercely loyal following in the process. From rural Japanese roads to snowbound American mountain passes, from the world's most grueling rally stages to the bustling heart of Tokyo, Subaru vehicles have been recognized by a signature blend of practicality and innovation—traits rooted in the company's earliest ambitions.

That origin story begins not on the ground, but in the air. The 1917 founding of the Nakajima Aircraft Company, the forerunner of Fuji Heavy Industries, laid a foundation built on precision, resilience, and a fascination with the physics of movement. In the aftermath of the Second World War, Japan's devastated economy forced companies like Nakajima to transform. Out of necessity and vision, a new identity emerged. Fuji Heavy Industries was born, and with it, the seeds of a car company that would soon challenge the conventions of an industry just finding its global feet.

Subaru's first car, the 360, launched in 1958, was a direct response to the Japanese government's call for practical, efficient vehicles to mobilize the masses. This compact, egg-shaped car was both a feat of engineering and a social statement, signaling that reliable mobility could be within reach for every family. It was a car designed less to impress than to serve—lightweight, frugal, and sturdy—and its success set the tone for a company that would never lose sight of its customers' real-world needs.

As Subaru grew, so did its reputation for charting an independent course. In the 1960s, while competitors chased bigger engines and flashier styling, Subaru engineers quietly adopted a mechanical layout that would become their signature: the horizontally opposed boxer engine. Paired with the development of affordable all-wheel drive—a feature once reserved for rugged trucks and military vehicles—Subaru created cars uniquely suited to challenging weather, rough terrain, and the everyday unpredictability of life on the road. These decisions were not born of marketing bravado, but of an engineering culture steeped in pragmatism and a willingness to try what others deemed unorthodox.

Subaru's practical genius found eager audiences in places as disparate as the rice paddies of Hokkaido and the snowbound towns of Vermont and Colorado. Throughout the 1970s and 1980s, Subaru's reputation for durability and all-weather capability cemented its status as the go-to choice for farmers, teachers, doctors, and adventurers alike. Models like the Leone, the BRAT, and the Legacy each carried forward the company's founding values, introducing innovations such as part-time and later full-time all-wheel drive to a world beginning to understand the importance of stability and safety in everyday driving.

Yet, Subaru's story is not one of simple functionality. The company's DNA, shaped by its aviation roots, is evident in the relentless pursuit of technical excellence. Subaru became a pioneer in turbocharging, environmental responsibility, and crash safety, and it embraced the challenge of rally racing not as a marketing exercise, but as a proving ground for its core technologies. Victories on the world stage—first in the hands of legends like Colin McRae, and later through successive generations of WRX and STI models—did more than sell cars. They forged a legend, binding together engineers, dealers, drivers, and fans in a shared story of resilience and performance.

It is this fusion of humility and heroism, of practicality and passion, that has made Subaru a phenomenon transcending markets and generations. The company's commitment to building cars that serve real needs has only strengthened over time, even as the world around it has changed. Subaru's environmental leadership, commitment to safety, and ability to foster loyalty among both rural families and urban enthusiasts has helped it weather crises, adapt to new technologies, and remain relevant as the definition of "car" evolves.

Subaru's enduring legacy is perhaps best understood not through its marketing slogans, but in the stories told by those who rely on its cars every day. For some, it is the memory of a first family trip in a battered 360, for others a WRX devouring gravel roads in the hands of a weekend rally driver, or an Outback carrying medical supplies across a snow-locked mountain pass. In these stories, Subaru is more than a manufacturer; it is a companion, a problem-solver, and sometimes, a lifeline.

As the chapters ahead trace Subaru's evolution—from the earliest days of Fuji Heavy Industries, through iconic models, motorsport triumphs, and global expansion—one thing becomes clear: Subaru's story is not defined by size, spectacle, or sheer numbers. It is a story of focus, of finding strength in specialization, and of building trust one driver at a time. The road from aircraft assembly lines to rally stages and city streets is long and winding, but every turn reveals a company unafraid to pursue its own vision.

This book is an invitation to journey along that road, to discover how a company forged in the aftermath of war became a symbol of resilience and innovation. It is a story written in snowdrifts and streetlights, on racetracks and remote dirt roads, in the everyday triumphs of people who simply need to get where they're going, no matter the conditions. Subaru's legacy, now more than ever, is a reminder that greatness sometimes comes not from following the crowd, but from forging ahead—no matter how rough the terrain.

Chapter 1: Wings and Wheels: Subaru's Aviation Origins

Subaru's journey as a car manufacturer cannot be understood without first exploring the turbulence and transformation that marked twentieth-century Japan. Long before the brand's blue-and-gold rally cars roared across world stages, its story began in the clouds, with a company that looked skyward for both inspiration and survival. The path to becoming a global automotive icon started in the hangars and workshops of the Nakajima Aircraft Company—a name synonymous with both technological prowess and the daunting challenges of war and reconstruction.

Founded in 1917 by Chikuhei Nakajima, a naval engineer with a vision for Japan's future in aviation, Nakajima Aircraft Company emerged as one of the country's most important manufacturers in the years leading up to and during the Second World War. With the burgeoning global fascination for flight and the strategic needs of a nation intent on rapid modernization, Nakajima Aircraft Company grew quickly, attracting some of Japan's brightest engineering talent. The company's early successes included a series of military and civilian aircraft, earning it a reputation for technical ingenuity, reliability, and the capacity to innovate under pressure.

As the 1930s unfolded, Nakajima's role in Japan's expanding military-industrial complex became ever more pronounced. Its factories, located in Ota City and elsewhere, became hubs of technological development, producing thousands of aircraft that would serve throughout the conflict. By 1944, Nakajima had become the largest aircraft manufacturer in Japan, supplying fighters such as the Ki-43 "Oscar" and the Ki-84 "Frank" to the Imperial Japanese Army Air Service. The company's engineers worked tirelessly to balance the competing demands of speed, durability, and performance—a balancing act that would later echo through every Subaru vehicle.

But the collapse of Imperial Japan in 1945 brought a dramatic halt to these operations. Under the terms of the Allied Occupation, Japan's once-mighty armaments industry was dismantled, and Nakajima Aircraft Company, like many other defense contractors, faced an uncertain future. Out of necessity and determination, the company splintered, with its core engineers and leadership seeking ways to repurpose their skills for peacetime production. The scars of war, however, would prove to be fertile ground for reinvention.

The legacy of Nakajima did not disappear; instead, it fragmented into several companies, each focusing on new industries. One of the most significant of these was Fuji Sangyo, founded in 1946. Fuji Sangyo gathered many of the same engineers who had cut their teeth on aircraft design, their expertise in light metals, precision assembly, and aerodynamics setting them apart from other postwar enterprises. At first, the company diversified into a variety of products—gasoline-powered bicycles, motor scooters, and even kitchen equipment—seeking any foothold in a devastated economy.

In this period of hardship, the seeds of what would become Subaru were quietly being sown. The need for affordable, practical transportation was immense, both in the cities and rural regions where Japan's infrastructure had been shattered. The engineers at Fuji Sangyo, drawing on their experience with aircraft, recognized the potential of lightweight, strong structures and the importance of reliable mechanical systems. Their earliest automotive efforts—experimental prototypes and microcars—reflected these values. Although few of these vehicles made it to large-scale production, the experience proved invaluable, fostering a culture of technical ingenuity and problem-solving that would become a Subaru hallmark.

The next pivotal chapter arrived with the corporate reorganization of 1950. Amidst new government regulations aimed at consolidating Japan's fractured industrial base, Fuji Sangyo was divided into a handful of smaller companies, each focused on a specific business sector. Yet the bonds formed among the aircraft engineers remained strong. Over the next several years, these entities—Fuji Kogyo (machinery), Fuji Jidosha (automotive), Omiya Fuji Kogyo (coachbuilding), Utsunomiya Sharyo (railways), and Tokyo Fuji Sangyo (sales and distribution)—would lay the groundwork for a new kind of industrial partnership.

By 1953, these companies recognized their shared heritage and complementary skills. Driven by a desire to reestablish themselves as a force in Japanese manufacturing, they merged to form Fuji Heavy Industries Ltd. The new conglomerate brought together expertise in metallurgy, assembly, logistics, and product design, all under the stewardship of men who had spent their formative years building aircraft for the most demanding of environments. The Fuji Heavy Industries name signaled both continuity with the past and a commitment to the future—a future in which flight-inspired innovation would find a new home on the road.

Crucially, the technical DNA of the new company was inextricably linked to its aviation past. Aircraft engineering had taught the Fuji team the value of lightness without fragility, of aerodynamics not merely for speed but for efficiency, and of systems designed for reliability in harsh, unpredictable conditions. These principles would find new life in every automotive project the company undertook. The emphasis on weight reduction, structural rigidity, and meticulous assembly carried over into Fuji Heavy Industries' earliest car designs.

One of the first and most significant peacetime products was the Fuji Rabbit, a motor scooter introduced in 1946. While not a car, the Rabbit was a revelation in postwar Japan: light, sturdy, and capable of navigating the rough, unpaved streets that still crisscrossed the country. Its success was directly attributable to the aircraft-derived techniques employed in its manufacture. Aluminum components, pressed steel frames, and compact, efficient engines set the Rabbit apart from other scooters, establishing Fuji Heavy Industries as a technological leader even in unfamiliar markets. The Rabbit would become one of the most popular scooters in Japan, produced in numerous variants and selling hundreds of thousands of units throughout its long production run.

The process of moving from two wheels to four, however, was not immediate. The immediate postwar years in Japan were marked by economic hardship, material shortages, and shifting consumer priorities. For the engineers and executives at Fuji Heavy Industries, the question was not merely how to build a car, but how to build one that was both technically sophisticated and accessible to the millions of Japanese families yearning for mobility. The answer lay in blending the best of their aircraft legacy—precision, innovation, resilience—with the hard realities of the automotive market.

It was in this crucible of necessity and ambition that the Subaru name was chosen. "Subaru" is the Japanese word for the Pleiades star cluster, a formation visible in the night sky and an ancient symbol of unity. The company's new logo, depicting six stars—one large, representing Fuji Heavy Industries, and five smaller, for the companies that had merged to form it—evoked both the celestial and the communal. The choice reflected a deeply held belief that the whole could be greater than the sum of its parts, and that teamwork, innovation, and persistence would light the path forward.

In developing their first automobile, the team at Fuji Heavy Industries drew directly from their past. Their approach to materials—favoring lightweight steel, robust chassis designs, and compact, efficient engines—was shaped by aircraft practices. Their insistence on rigorous testing, precision tolerances, and quality control was not merely a matter of pride but of survival, given the harsh scrutiny and fierce competition of the emerging Japanese car market. Every design decision was weighed against the twin standards of durability and practicality, echoing the realities of wartime aircraft manufacturing where mistakes could not be afforded.

The Japanese Ministry of International Trade and Industry (MITI), which played a pivotal role in guiding industrial policy during this period, launched the "People's Car" program in the mid-1950s. Its aim was to spur the creation of a small, affordable vehicle capable of mobilizing the nation. Fuji Heavy Industries was among the companies inspired by this initiative, setting the stage for a surge of creativity and competition that would soon revolutionize Japanese motoring.

While other manufacturers experimented with various layouts and engineering concepts, the Subaru team's collective memory of aircraft production guided them toward innovative solutions. They prioritized mechanical simplicity, ease of maintenance, and resilience in adverse conditions. These characteristics would become the bedrock of the Subaru reputation in the decades to follow, influencing everything from engine layout to suspension design.

As the 1950s drew to a close, the first Subaru car—the 360—was readied for production. But even before that milestone, the groundwork had been laid. Fuji Heavy Industries had transformed itself from a maker of warplanes to a builder of vehicles for peace, drawing on the best of its heritage while meeting the urgent needs of a changing society. Its engineers had transferred the lessons of flight to the challenges of ground travel, creating a technical culture that valued innovation for its own sake and always looked for new ways to serve the community.

In the years to come, Subaru's name would be inextricably linked to ruggedness, ingenuity, and a willingness to go where others would not. The aircraft engineers of Nakajima and Fuji had set the standard: build vehicles that last, vehicles that inspire confidence, and vehicles that carry forward a spirit of invention born in the shadow of adversity. The path from wings to wheels was neither straight nor smooth, but it was defined by the same unyielding pursuit of progress that had once sent Japanese aircraft into the skies.

From these early years, a philosophy emerged that continues to shape Subaru's approach: look to the future, respect the lessons of the past, and never lose sight of the people who trust your machines with their lives. The merging of aviation excellence and automotive ambition gave rise to a company ready to meet the challenges of a dynamic and demanding world. As the Subaru badge began to appear on city streets and country roads, its quiet confidence and technical distinctiveness spoke volumes about the journey from propellers to pistons, from aircraft hangars to family driveways.

The roots of Subaru are deeply entwined with the history of Japanese innovation and resilience. Its transformation from the Nakajima Aircraft Company to Fuji Heavy Industries, and finally to a pioneering automaker, laid a foundation for decades of engineering achievement. The spirit of those first engineers—resourceful, adaptable, and relentless in the face of adversity—remains embedded in every Subaru that follows, setting the stage for a brand that would soon become synonymous with practical excellence and a quiet, enduring form of greatness.

Chapter 2: The Rabbit Emerges: Launching the Subaru 360

In the late 1950s, Japan stood at the threshold of a transformation that would remake its cities, its industries, and the daily lives of its people. Recovery from the devastation of the Second World War had demanded sacrifice, resourcefulness, and a collective effort to build anew. As the economy began to stir, the dream of personal mobility—once reserved for the privileged—spread rapidly among a population eager for change. It was into this moment, defined by optimism and necessity, that Subaru would introduce the car that would make its name: the Subaru 360.

The story of the 360's creation begins with the bold ambitions of Fuji Heavy Industries, a company reborn from the engineering expertise of the Nakajima Aircraft Company and guided by a vision to put Japan on wheels. With the success of the Fuji Rabbit scooter already proving that lightweight, durable vehicles could thrive in a recovering nation, Fuji Heavy Industries set its sights on an even greater challenge: building a car that was affordable, efficient, and suitable for millions of ordinary Japanese citizens.

This challenge found its focus in the "People's Car" initiative, a government-backed effort to encourage the mass production of compact, low-cost vehicles that could spur economic growth and transform the nation's infrastructure. In 1955, the Ministry of International Trade and Industry (MITI) outlined a set of specifications for what this new car should be: a four-seater, under 360cc displacement, weighing less than 360 kilograms, and capable of at least 100 kilometers per hour. Meeting these requirements demanded technical ingenuity and a deep understanding of the needs of Japanese families, who navigated narrow streets and tight budgets.

Leading the project for Fuji Heavy Industries was engineer Kenji Kita, a man with a passionate belief that building a car was not simply about assembling metal and parts, but about capturing the spirit of progress. Kita's team, comprised largely of veterans from the company's aircraft days, approached the problem with the precision and creativity that had defined their wartime efforts. Their challenge was formidable: design a car from scratch with limited resources, scarce materials, and an unproven market for such a vehicle.

After several prototypes and a rigorous process of trial and error, the result was unveiled in 1957—the Subaru 360. The car made its official debut at the Tokyo Motor Show, where it stood out for its unusual, rounded shape and almost toy-like proportions. Measuring just three meters long, with a width of 1.3 meters and a height of 1.4 meters, the Subaru 360's compactness was both a technical solution and a statement of purpose. Every line and curve was dictated by a drive for efficiency, lightness, and maximized interior space.

At the heart of the 360 was a two-cylinder, air-cooled, 356cc engine, mounted in the rear and producing a modest 16 horsepower. The engine was designed with an emphasis on simplicity and ease of maintenance, with parts that could be manufactured and serviced using existing Japanese industry capabilities. Weighing only 385 kilograms, the car was remarkably light, a testament to the team's experience with aircraft construction and their careful selection of materials, including thin-gauge steel and lightweight plastics.

The car's distinctive "monocoque" body construction, rare at the time in Japan, provided structural rigidity without adding excess weight. Suspension was handled by independent coil springs at the front and a torsion bar setup at the rear, offering a surprisingly comfortable ride for such a small vehicle. The interior, though basic, was thoughtfully arranged to maximize the sense of space, and the rear-engine layout allowed for a flat floor that enhanced passenger comfort.

Upon its launch in 1958, the Subaru 360 quickly captured public attention. The car's rounded, almost beetle-like appearance led to affectionate nicknames such as "ladybug" ("tentomushi" in Japanese), and its modest price tag put it within reach of many first-time car buyers. With a starting price of 425,000 yen—about half the average annual salary in Japan at the time—the 360 represented a genuine breakthrough in accessibility. Families who had previously relied on bicycles or scooters could now imagine owning a car.

The reception was immediate and enthusiastic. Dealers struggled to keep up with demand as customers flocked to see the new vehicle that promised independence and convenience. The 360's performance, while modest by Western standards, was perfectly attuned to the realities of Japanese roads, which were narrow, congested, and often unpaved. Capable of reaching 80 kilometers per hour, with fuel efficiency that routinely exceeded 30 kilometers per liter, the Subaru 360 was exactly the right car at exactly the right time.

Its compact size was more than a convenience—it was a necessity. Japanese cities were filled with tight alleys, small parking spaces, and crowded neighborhoods where larger cars would simply not fit. The 360's nimbleness and small turning radius made it ideal for urban driving, while its sturdy construction allowed it to withstand the rougher conditions of rural roads. This versatility was crucial to its widespread adoption, as it suited the needs of both city dwellers and countryside families.

The success of the 360 had far-reaching implications for both Fuji Heavy Industries and the Japanese automotive industry as a whole. For Subaru, it was a validation of its approach: prioritize the needs of ordinary people, leverage technical ingenuity, and never lose sight of reliability and practicality. For Japan, it marked the true beginning of mass motorization. Sales of the 360 soared, with production passing 10,000 units in the first year and climbing steadily thereafter. By the end of its 12-year production run in 1971, more than 392,000 units had been sold—a remarkable achievement for a first attempt at car manufacturing.

Beyond sales numbers, the 360 established key precedents for Subaru's corporate identity. The focus on lightweight design, efficient engineering, and value for money would become hallmarks of the brand for decades to come. The use of unconventional solutions—such as the rear-mounted engine and monocoque body—demonstrated a willingness to challenge norms and seek better answers, even when it meant taking risks.

The Subaru 360's impact was also felt in Japanese society at large. It became a symbol of upward mobility and modernity, featuring in advertisements that highlighted young families, adventurous outings, and the newfound freedom of weekend getaways. The car's approachable image helped dissolve lingering notions that cars were only for the elite or the urban professional. Now, shopkeepers, farmers, teachers, and factory workers could all aspire to car ownership.

The press responded with fascination and, often, admiration. Automotive journalists noted the 360's creative solutions to technical challenges and its role in democratizing transportation. International observers were more skeptical at first, sometimes mocking its tiny size and modest power. Yet, over time, even foreign critics would come to appreciate the car's ingenuity, especially as Japan's economic resurgence and manufacturing prowess began to attract worldwide attention.

In engineering circles, the 360's development was a case study in making the most of constraints. Each challenge, from material shortages to budget limitations, was met with resourcefulness and a clear-eyed assessment of priorities. The success of the project validated the principle that necessity could drive innovation, and that simplicity—when thoughtfully executed—could deliver not just cost savings but real, everyday value for users.

The 360's influence rippled beyond its own production line. Its success encouraged Fuji Heavy Industries to continue developing new automotive models, leading to a succession of vehicles that built on the lessons learned. It inspired competing manufacturers to invest more heavily in compact car design, sparking an era of innovation that would propel Japanese cars onto the global stage within a decade.

The 360 was also significant for what it represented internally at Fuji Heavy Industries. It brought together engineers from varied backgrounds—aviation, motorcycles, heavy equipment—in a common purpose. The project fostered a sense of teamwork and a culture of creative problem-solving that would serve the company well in years to come. The pride of building Japan's "people's car" was evident in every department, from design studios to assembly lines.

As with any first, the Subaru 360 was not without its challenges. Early models experienced teething problems, particularly with reliability under prolonged use or in harsher environments. Fuji Heavy Industries responded quickly, introducing improvements and listening closely to customer feedback. This willingness to adapt and improve became another cornerstone of Subaru's approach—an acknowledgment that the relationship between automaker and driver did not end at the point of sale.

The 360's story was also about scale. For the first time, Fuji Heavy Industries had to organize mass production on a level far beyond anything it had attempted before. New manufacturing techniques were developed, supply chains established, and dealer networks expanded to reach customers across the length and breadth of Japan. The experience gained in scaling up production would prove invaluable as Subaru prepared to launch future models both at home and, eventually, abroad.

In retrospect, the Subaru 360 was more than just a car. It was a declaration of intent—a statement that Fuji Heavy Industries would be a company defined by its ability to meet the real needs of ordinary people with ingenuity, integrity, and perseverance. It set the precedent for accessible, reliable transport that would become synonymous with the Subaru name. For thousands of Japanese families, it was the first step into a wider world, a vehicle that carried hopes as well as passengers.

In the years that followed, the 360's DNA would be visible in every Subaru: a focus on practical design, a readiness to defy convention, and an unyielding commitment to reliability. The launch of the Subaru 360 did more than introduce a new car—it established a new way of thinking about mobility in Japan.

The "ladybug" became an icon, but more importantly, it became a tool for transformation, helping to drive the country forward, one modest, dependable journey at a time.

Chapter 3: Building a Brand: The 1960s and Subaru's Identity

As Japan emerged from the hardships of the postwar period into the energetic promise of the 1960s, a new spirit of optimism coursed through its cities and countryside alike. The echoes of economic revival could be heard in the clatter of factory assembly lines, the buzz of scooters in narrow alleyways, and the rising hum of compact cars on roads that grew busier by the day. At the heart of this transformation stood a modest but increasingly ambitious automaker: Subaru. Having made its mark with the 360, Subaru faced the pivotal challenge of turning a single model's success into a sustainable, recognizable brand identity—a task that would define its destiny throughout the decade.

The 1960s opened with the 360 still at the center of Subaru's efforts, its continued popularity fueling the company's first meaningful profits and providing the foundation for investment in new ventures. Yet, as Japanese society changed, so too did its needs. Rapid urbanization created new demands for nimble, reliable vehicles, while the countryside remained a stronghold for practical, versatile transport. Recognizing these trends, Fuji Heavy Industries committed to broadening its lineup, determined to keep pace with a population on the move.

A major step came in 1961 with the introduction of the Subaru Sambar. At first glance, the Sambar was an unassuming microvan, but its significance for Subaru's future was immense. Drawing on lessons learned from the 360's lightweight construction and efficient packaging, the Sambar offered a clever adaptation for commercial and family use. Built on a rear-engine, rear-wheel-drive layout, the Sambar employed the same 356cc two-stroke engine that powered the 360, but packaged it in a compact, boxy body capable of carrying goods, equipment, and up to four passengers.

The Sambar's arrival met an enthusiastic reception among small business owners, farmers, and tradespeople, who prized its maneuverability, low operating costs, and ability to squeeze through narrow city lanes or rough country paths. Its flat cargo floor, independent suspension, and simple maintenance needs made it a workhorse for a rapidly modernizing Japan. Over time, the Sambar's versatility gave rise to a host of variants, including pickup trucks and specialized commercial models, cementing its reputation as the practical companion to the 360's family-friendly image.

In the same period, Subaru continued refining its core offerings. The 360 saw continuous improvements in comfort, reliability, and style, as engineers responded to customer feedback and the evolving expectations of Japanese motorists. Innovations such as better weatherproofing, enhanced seating, and more robust engines kept the model competitive, even as rival automakers flooded the market with their own compact cars. Fuji Heavy Industries paid careful attention to detail, focusing on user-friendly features and durability, characteristics that would soon become synonymous with the Subaru name.

The company's strategy was clear: rather than chase fleeting trends or rapid model churn, Subaru focused on incremental progress, technical ingenuity, and building trust with its customers. This approach paid dividends as the 1960s wore on. The Sambar's reliability in rural areas helped solidify Subaru's appeal outside of urban centers, while the 360's gentle, approachable character earned the affection of city families and first-time drivers. Word of mouth and positive ownership experiences became invaluable marketing tools, fostering a sense of community around the brand.

As its reputation grew, Subaru began to think beyond domestic borders. While the Japanese market was its lifeblood, the company recognized the long-term necessity of international expansion—a challenge that required careful preparation and a distinct identity. Early forays into export markets were tentative, with small numbers of 360s and Sambars shipped to select destinations in Southeast Asia, Australia, and a handful of European countries. The results were mixed: some markets embraced the quirky, efficient vehicles, while others, accustomed to larger, more powerful cars, greeted them with skepticism or amusement.

Still, these early exports taught Fuji Heavy Industries vital lessons about foreign tastes and standards. The need to adapt to different regulations, climates, and consumer preferences prompted refinements in engineering and manufacturing, setting the stage for more serious export efforts in the years to follow. This process also forced Subaru to define what made its cars unique—what it could offer that competitors could not.

In the midst of these ambitions, Fuji Heavy Industries began experimenting with new models that would further shape Subaru's identity. In 1966, the Subaru 1000 was introduced, marking a dramatic leap forward in both engineering and brand philosophy. The 1000 departed from the rear-engine layout of earlier models and instead featured a water-cooled, four-cylinder, horizontally opposed (boxer) engine mounted at the front, paired with front-wheel drive. This combination was a technical milestone—not just for Subaru, but for the Japanese auto industry as a whole.

The horizontally opposed engine, more commonly seen in European sports cars than in Japanese family sedans, offered inherent balance and smoothness, lowering the car's center of gravity and improving handling. Paired with front-wheel drive, the Subaru 1000 delivered exceptional traction and stability, especially on slippery or uneven roads. This innovation set Subaru apart, signaling the company's willingness to chart its own course rather than follow the conventions of its domestic rivals. The 1000 also introduced improved safety features, such as dual-circuit brakes, and a roomier, more comfortable cabin—a nod to the growing expectations of Japanese drivers as living standards rose.

The launch of the 1000 marked the birth of a new technical identity for Subaru, one centered around engineering excellence, unconventional solutions, and a refusal to compromise on reliability. The model's success, though not overwhelming in terms of sales volume, gave Subaru the confidence to pursue technical distinctiveness as a long-term strategy. The boxer engine, in particular, would become a Subaru signature, woven into the DNA of almost every model for decades to come.

Throughout the decade, Subaru's public image was carefully nurtured through advertising campaigns that emphasized practicality, safety, and the everyday adventures of ordinary people. Print ads and posters featured cheerful families and hard-working tradesmen, reinforcing the idea that Subaru was a brand for those who valued dependability and honesty. The brand's logo, representing the Pleiades star cluster, grew increasingly familiar on city streets and country lanes, a quiet symbol of unity and forward motion.

Dealers and service networks expanded in step with Subaru's growing lineup. Unlike some competitors who prioritized rapid growth at any cost, Fuji Heavy Industries built relationships with trusted dealers, insisting on high standards of service and customer care. This approach, though slower, paid off in lasting customer loyalty and a reputation for after-sales support. Subaru's emphasis on listening to owners and incorporating their feedback further strengthened this bond, helping to turn first-time buyers into lifelong advocates.

By the close of the 1960s, Subaru's place in the Japanese automotive landscape was secure. Its models, though rarely the flashiest or fastest, were prized for their ability to withstand the challenges of everyday life. The Sambar and the 360 had both become staples on city streets and rural backroads, serving everyone from shopkeepers to farmers, students to grandparents. The Subaru 1000 had planted the seeds of technical innovation that would soon blossom into one of the industry's most distinctive engineering philosophies.

Just as importantly, Subaru had established the core values that would guide its evolution: a focus on practical, user-oriented design; a culture of technical creativity rooted in its aviation heritage; and a belief that reliability and honesty would win the trust of drivers everywhere. As competitors jockeyed for attention with ever-larger lineups and aggressive marketing, Subaru quietly built a brand on the strength of its actions and the satisfaction of its customers.

The 1960s had proven that Subaru could do more than survive—they could innovate, differentiate, and lay the groundwork for a future defined by technical distinction and unshakable dependability. The path ahead would bring new challenges and greater ambitions, but the foundation had been set: Subaru was no longer just a maker of clever little cars, but a company with an identity, a purpose, and a growing following at home and, soon, around the world.

Chapter 4: Engineering Distinction: Boxer Engines and Technical Innovation

In the panorama of global automotive engineering, certain technical choices define not just models, but entire brands. For Subaru, that defining choice came in the form of the horizontally opposed engine, a mechanical solution that, while unconventional in its home market, would set the company apart from its earliest years as a carmaker and become central to its enduring identity. As the 1960s gave way to a new era of innovation and ambition, the boxer engine emerged not merely as a technical flourish, but as a statement of purpose—one that would shape Subaru's vehicles, reputation, and future.

The horizontally opposed engine, often called a "boxer" because of the way its pistons move in pairs, punching outward like the gloved fists of sparring fighters, was not an invention of Subaru's own. Its origins stretched back to early twentieth-century Europe, where German and British engineers explored its advantages for both cars and aircraft. Porsche, Volkswagen, and certain high-performance aircraft manufacturers had long touted the configuration's inherent smoothness, compactness, and natural balance. Yet, in the Japan of the mid-1960s, it was still a rarity, especially in the mass market.

Subaru's adoption of the boxer engine was rooted in both technical necessity and creative vision. As Fuji Heavy Industries prepared to expand its automotive range beyond the rear-engined 360 and Sambar, its engineering team faced a pivotal question: how to deliver improved comfort, performance, and safety for a new generation of drivers, without abandoning the company's reputation for compact, accessible vehicles? The answer would draw as much from the company's aviation legacy as from automotive precedent.

The launch of the Subaru 1000 in 1966 marked a watershed. It was the first Japanese car to feature both a water-cooled, horizontally opposed four-cylinder engine and front-wheel drive—a pairing that would come to define much of Subaru's later lineup. The choice was not arbitrary. Engineer Shinroku Momose, who had previously worked on fighter aircraft at Nakajima, championed the boxer configuration for its balance and low center of gravity. In a conventional inline engine, the vertical stacking of cylinders raises the vehicle's mass, increasing the risk of body roll and reducing agility. By contrast, the flat profile of the boxer engine allowed it to be mounted lower in the chassis, distributing weight more evenly and improving handling, especially in emergency maneuvers or on challenging road surfaces.

The technical merits of the boxer engine extended beyond stability. Its opposed pistons moved in perfect counterpoint, naturally cancelling out each other's vibrations and reducing engine harshness. The result was a smooth-running powerplant with less need for complex counterbalancing or heavy engine mounts. For everyday drivers, this translated into a quieter, more refined ride—an important selling point in an era when Japanese cars were still fighting perceptions of roughness and austerity.

The compactness of the engine also afforded new freedoms in vehicle design. With less vertical bulk and a shorter length compared to equivalent inline-four engines, the boxer allowed Subaru to craft lower, sleeker hoods and maximize interior space. In the Subaru 1000, the flat-four's placement ahead of the front axle facilitated a roomy cabin with minimal intrusion from drivetrain components, a critical advantage as

Japanese families demanded more comfort and utility from their cars.

Just as important as the engine itself was its partnership with front-wheel drive. While most domestic competitors still relied on rear-wheel-drive layouts, Fuji Heavy Industries' engineers recognized that pairing the boxer engine with front-wheel drive offered numerous advantages for traction, especially in the wet or on loose surfaces. The weight of the engine pressing down over the driven wheels improved grip, while the car's inherently balanced chassis inspired confidence in all conditions. This combination would prove prescient, particularly as Subaru's ambitions in all-weather capability and, later, all-wheel drive began to take shape.

The boxer engine's debut in the Subaru 1000 was the product of painstaking development and rigorous testing. Early prototypes underwent thousands of kilometers of road and laboratory evaluation, as engineers tweaked everything from cooling systems to crankshaft design in pursuit of reliability and efficiency. The result was a 977cc engine delivering 55 horsepower—modest by today's standards, but lively and flexible for its day. The car's performance, especially in comparison to its size and class, was praised for both its agility and composure, thanks in no small part to the engine's characteristics.

The public and press quickly noticed the Subaru 1000's difference. Reviews highlighted the car's stability at speed, its surprisingly nimble handling, and the refined hum of its engine compared to the harsher notes of competitors' inline-threes and -fours. As more families took delivery of the 1000 and experienced the benefits first-hand, word of mouth added to the growing sense that Subaru was a company willing to innovate, not just imitate.

The boxer engine soon became a foundation on which Subaru's reputation for durability and dependability would be built. Its robust construction, straightforward maintenance, and cool-running design lent itself well to the varied demands of Japanese and, increasingly, overseas drivers. Mechanics appreciated the engine's accessibility and logical layout; owners enjoyed its long service life and consistent performance, even in demanding environments.

Encouraged by this positive reception, Subaru doubled down on its technical approach. Successors to the 1000, including the FF-1 Star and later the Leone, carried forward the boxer layout, refining the design with new displacements, improved carburetion, and, eventually, the introduction of overhead camshafts. By the early 1970s, Subaru was synonymous with the flat-four, a rare feat for a manufacturer in the fiercely competitive world of mass-market cars.

The benefits of this commitment extended beyond technical satisfaction. The boxer engine became a visible, even marketable, point of difference in an increasingly crowded field. Advertisements and brochures highlighted the unique engine layout, inviting buyers to "feel the balance" and "experience the smoothness" that set Subaru apart. For the mechanically inclined, it was a sign of thoughtful engineering; for the everyday driver, it was a quiet assurance that Subaru did things differently, and often better, than the rest.

Subaru's willingness to invest in the boxer platform paid dividends as the company's ambitions grew. The engine proved adaptable to a range of applications, from compact sedans and wagons to, eventually, performance cars and four-wheel-drive models. Its inherent balance and low center of gravity made it ideal for rally racing, where stability and predictability on loose surfaces were crucial. As Subaru's motorsport story began to unfold in the decades that followed, the boxer engine would be at the heart of every victory, every innovation, and every legendary drive.

The technical distinctiveness also created a culture of pride within Fuji Heavy Industries. Engineers, dealers, and even marketing staff found a shared identity in the company's refusal to follow the crowd. The boxer engine was not the cheapest, easiest, or most conventional path—but it was the right one for Subaru's goals. The sense of mission, of building something unique and valuable, infused every department, reinforcing a corporate philosophy that prized long-term excellence over short-term trends.

As emissions regulations tightened in the 1970s and fuel economy became a paramount concern, Subaru's boxer engine evolved in step. Engineers developed new combustion chambers, revised cam profiles, and, in time, introduced electronic fuel injection to keep pace with changing standards. The engine's compact footprint made it easier to accommodate catalytic converters and other emissions-control devices without sacrificing performance or reliability. This adaptability would serve Subaru well as it expanded into export markets with differing regulatory demands.

For Subaru's growing base of loyal customers, the boxer engine was more than a technical curiosity. It became a symbol—a badge of belonging to a community that valued engineering honesty and tangible benefits over flash and fashion. Subaru owners, from the earliest 1000 drivers to future generations of WRX and Outback enthusiasts, took pride in knowing that what sat beneath their hood was a product of purposeful design, refined by decades of expertise and relentless testing.

The company's steadfast commitment to the boxer layout set the tone for decades of innovation to follow. In an age of consolidation and commoditization, Subaru maintained its dedication to doing things differently, earning respect from engineers, drivers, and competitors around the world. As the lineup expanded, as new models and new challenges emerged, the horizontally opposed engine remained at the core of Subaru's identity—a silent, powerful reminder that technical distinction can be the foundation of greatness.

By the end of the 1970s, Subaru's engineering philosophy was clear: embrace what works, refine what can be improved, and never hesitate to take the road less traveled if it leads to a better solution for drivers. The boxer engine, with its humble origins and remarkable adaptability, became the thread that wove together every chapter of Subaru's ongoing story, carrying forward a spirit of innovation, balance, and engineering courage that continues to define the brand to this day.

Chapter 5: All-Wheel Drive for All: The 4WD Revolution

By the late 1960s and early 1970s, Japan's automotive industry was a global force, and Subaru had earned a reputation for unconventional engineering and dependable small cars. Yet, it was in the area of drivetrain technology—specifically, the application of all-wheel drive (AWD) to passenger vehicles—that Subaru would make one of its most consequential and lasting contributions to automotive history. What began as a response to practical needs in rural Japan would eventually transform the company's fortunes, reshape customer expectations, and propel Subaru to the forefront of global AWD innovation.

The roots of Subaru's AWD journey lay in the everyday realities faced by Japanese drivers outside the nation's burgeoning cities. Rural Japan, with its winding mountain roads, sudden snowfalls, and frequent muddy or unpaved stretches, posed challenges that conventional rear- or front-wheel-drive vehicles could not always meet. Farmers, delivery workers, and utility companies needed reliable transportation capable of handling both paved and treacherous surfaces. While four-wheel-drive vehicles existed, they were typically built on truck or off-road platforms—rugged and capable, but unsuited to the comfort, size, and efficiency required by everyday motorists.

In 1971, Fuji Heavy Industries made a move that would change both its destiny and the landscape of passenger vehicles: the introduction of the Subaru Leone 4WD Station Wagon. While not the world's first four-wheel-drive car, it was the first mass-produced, affordable passenger wagon equipped with a selectable four-wheel-drive system, and crucially, it was designed for ordinary families, not for the military or extreme off-roaders.

The inspiration for this breakthrough came from a request by the Tohoku Electric Power Company, which needed vehicles able to reach remote transmission lines regardless of season or weather. Subaru engineers responded by adapting their proven Leone wagon with a new transfer case and drive system, enabling the driver to switch from two-wheel drive to four-wheel drive via a lever next to the gearshift. This relatively simple but robust system distributed engine power to all four wheels when needed, vastly improving traction on slippery or loose surfaces.

The debut of the Leone 4WD Wagon at the Tokyo Motor Show caused a stir among journalists and the public alike. Here was a car that combined the comfort, efficiency, and ride quality of a sedan with the sure-footedness of a truck—without the size, expense, or fuel consumption penalties of a traditional 4x4. The benefits were immediate and tangible. On snow, mud, or gravel, the Leone 4WD inspired a new level of confidence, not only for rural utility workers but also for families living in regions with harsh winters or unpredictable weather.

The early commercial success of the Leone 4WD Wagon in Japan led Fuji Heavy Industries to expand the offering across more of its product range. Other Leone body styles, including sedans and coupes, soon received the 4WD option. Dealers in the snowbelt regions of Japan, and later in countries with similar climates, found the vehicles in constant demand. The marketing slogan, "All-Wheel Drive for All," began to resonate with a public increasingly aware of the advantages of extra traction and stability.

Subaru's engineering team continued to refine its AWD systems through the 1970s and 1980s. Initial models used part-time, selectable four-wheel drive, but as consumer expectations and technical possibilities evolved, the company developed more sophisticated solutions. By the early 1980s, Subaru introduced a full-time four-wheel-drive system, employing a center differential that allowed continuous power delivery to all four wheels without driver intervention. This advance made AWD practical not just for slippery roads, but for every kind of driving—rain or shine, pavement or dirt, city or countryside.

These breakthroughs arrived just as all-weather safety and year-round usability were becoming central concerns for global motorists. In markets like the United States, where winter driving could be treacherous and rural areas expansive, Subaru's AWD vehicles found a ready audience. The Leone and its successors, such as the GL/DL wagon and the Justy, began to attract a following among drivers in New England, the Rocky Mountains, and the Pacific Northwest. Their combination of modest size, low operating costs, and reliable all-weather traction filled a niche that larger, thirstier SUVs and pickup trucks could not.

Subaru's commitment to AWD quickly became more than a technical feature—it became a cornerstone of the brand's identity and a key differentiator in an increasingly crowded marketplace. While most competitors offered AWD only as an expensive or limited option, Subaru steadily expanded the technology across its lineup, eventually making it standard on most models. By the late 1980s and early 1990s, all-wheel drive was no longer a niche selling point; it was central to Subaru's pitch and the reason many buyers chose the brand over others.

The benefits of Subaru's AWD systems were widely recognized not only by owners but by automotive journalists and safety experts. Comparative tests routinely found that

Subaru vehicles outperformed rivals in snow, rain, and on loose surfaces. In crash-avoidance and emergency-handling scenarios, the added grip of AWD contributed to shorter stopping distances, better control, and a general sense of confidence behind the wheel. Subaru leaned into these strengths with advertising campaigns that showed families traversing snowy mountain passes, city commuters braving rain-soaked streets, and outdoor enthusiasts exploring the limits of adventure.

Technically, Subaru's AWD systems evolved into several distinct types, including viscous-coupling center differentials, electronically controlled multi-plate clutches, and later, symmetrical full-time AWD layouts that optimized weight distribution and power delivery. The company's engineers worked tirelessly to ensure that the mechanical complexity of AWD did not compromise reliability or ease of maintenance. Service intervals, parts availability, and robust engineering made Subaru AWD vehicles as durable as their two-wheel-drive counterparts—a vital factor for owners in demanding climates and regions.

By the 1990s, Subaru's position as a leader in AWD technology was cemented on a global scale. In Japan, the brand became the default choice for drivers living in snowbelt regions or mountainous terrain. In North America, Subaru's reputation for all-weather competence found eager buyers among skiers, hikers, teachers, and rural professionals who needed reliable transport year-round. In Australia and Europe, Subaru's AWD cars became fixtures on country roads, city streets, and even rally circuits—where the advantages of constant traction were put to the ultimate test.

The impact of Subaru's AWD leadership extended beyond private owners. Police, emergency services, and utility companies in numerous countries adopted Subaru vehicles for their fleets, citing their unmatched ability to reach locations that would stop conventional cars. In Switzerland, Norway, and parts of the United States, Subaru wagons and sedans became a familiar sight wherever dependable winter mobility was essential.

All the while, Subaru's engineers continued to push the boundaries of AWD capability and refinement. In the early 1990s, the introduction of the Legacy and Impreza models brought advanced, symmetrical AWD systems that further enhanced handling and safety. These layouts featured a longitudinally mounted boxer engine and a straight, symmetrical drivetrain from front to rear axles, ensuring balanced weight distribution and minimizing torque steer. The technical harmony of the boxer engine and AWD system became a Subaru hallmark, providing a uniquely stable and predictable driving experience.

The adoption of AWD as a standard feature was not without its challenges. The added complexity and production costs required careful management, and Subaru's marketing team faced the task of explaining the benefits to customers who had never before considered AWD in a car. Yet, the company's commitment never wavered. The brand's message—confidence in all conditions, adventure without limits—resonated deeply, forging a connection with buyers who valued safety, practicality, and the ability to go farther.

In time, Subaru's pioneering approach would influence the entire industry. Today, the presence of AWD in everything from luxury sedans to compact crossovers owes much to the groundwork laid by Subaru's engineers in the 1970s and

1980s. The idea that all-wheel drive could be a tool for everyday confidence, not just off-road adventure, is now widely accepted—a testament to the foresight and persistence of Subaru's technical and product planners.

From a practical solution for Japanese utility workers to a global benchmark for all-weather driving, Subaru's AWD revolution fundamentally changed what customers could expect from a passenger car. The company's relentless pursuit of technical innovation, guided by the needs of real drivers on real roads, turned an engineering curiosity into a core brand promise. The path from niche option to global leader was built not on hype, but on results, and the millions of Subaru vehicles traversing snow, rain, gravel, and tarmac around the world are living proof of a legacy born from invention, perseverance, and a commitment to making every journey possible—no matter the conditions.

Chapter 6: Rural Roots: Subaru in the Japanese Countryside

Beneath the neon glow of Tokyo and far beyond the bustle of Osaka, the rural landscapes of Japan stretch across mountains, valleys, and coastlines in a tapestry of rice paddies, forested hills, and fishing villages. It was in these rural regions, defined by challenging terrain and unpredictable weather, that Subaru's reputation as a builder of reliable, rugged vehicles took deep root. While much of Japan's postwar economic miracle was played out in cities, the needs and rhythms of the countryside shaped both the direction and identity of Subaru, turning the brand into a trusted companion for farmers, tradespeople, and entire communities living at the country's margins.

The story of Subaru's rural popularity began with practical necessity. In the decades following World War II, Japan's rural areas faced unique obstacles: narrow, unpaved roads; harsh winters with heavy snowfall; steep, twisting mountain paths; and limited access to advanced infrastructure. Daily life required tools and vehicles that could withstand these challenges without faltering. For many families and small businesses, a car was more than a means of transport—it was an essential lifeline, a carrier of produce, children, supplies, and sometimes hope itself.

The earliest Subaru vehicles, particularly the 360 and Sambar, arrived in this context. The Sambar microvan and pickup, launched in 1961, quickly gained a following among rural drivers. Its compact size and nimble handling allowed it to traverse narrow farm lanes and village streets where larger trucks could not go. The Sambar's rear-engine layout, flat cargo floor, and robust suspension made it uniquely suited to carrying everything from rice sacks to building materials, while the modest yet dependable two-stroke engine proved easy to maintain in remote settings. It became common to see Sambars shuttling between fields and markets, or parked outside country workshops, its boxy silhouette a familiar part of the rural landscape.

The Subaru 360, though initially designed as a city car, found an unexpected home in rural Japan as well. Its light weight, simple mechanics, and fuel efficiency made it a practical choice for families on tight budgets. Farmers used the 360 for errands and market trips, appreciating the car's ability to handle rutted roads and its knack for starting reliably even on cold mornings. Its modest engine was untroubled by the frequent short journeys that characterized rural life, and its compactness allowed it to park in tight spaces alongside traditional farm equipment.

As the 1960s gave way to the 1970s, Subaru's commitment to rural customers deepened with the introduction of all-wheel drive—a breakthrough that would set the brand apart for generations. The Leone 4WD Station Wagon, launched in 1971, was an instant success among rural drivers. For the first time, farmers and villagers had access to a comfortable, affordable passenger vehicle capable of handling the mud, snow, and steep grades that defined so many country roads. The selectable four-wheel-drive system offered confidence without sacrificing fuel economy or day-to-day usability. From Hokkaido's snowy expanses to the winding tracks of Kyushu, the Leone 4WD became a mainstay of rural mobility.

Dealership networks played a crucial role in Subaru's rural ascendancy. Recognizing that urban-focused sales strategies would leave large swathes of the population underserved, Fuji Heavy Industries worked diligently to establish and support dealerships in small towns and rural prefectures.

Local dealers became more than just sales agents—they were community fixtures, offering advice, maintenance, and emergency repairs to neighbors and friends. This personal, service-oriented approach fostered deep loyalty and trust, qualities that urban-focused competitors sometimes struggled to match.

Reliability became the cornerstone of Subaru's rural reputation. Word spread quickly in tight-knit farming communities: a Subaru would start on the coldest morning, haul a heavy load up a muddy slope, and carry children safely to school even when other vehicles were stranded. Farmers told stories of their Sambars and Leones surviving years of rough use with minimal repairs, while shopkeepers praised the ability to quickly source replacement parts or perform basic fixes themselves. For many, a Subaru was less a product and more an extension of daily life—dependable, unpretentious, and always ready for work.

Seasonal extremes amplified the need for such reliability. In regions like Nagano, Niigata, and Akita, deep snowdrifts and icy roads made winter travel treacherous. Subaru's all-wheel-drive wagons, with their excellent traction and balanced handling, allowed rural families to travel with a measure of safety unknown in the days of rear-wheel-drive sedans or pickup trucks. In spring and autumn, when typhoons brought floods and washed out roads, the ruggedness of Subaru's suspension and chassis often made the difference between reaching home and being stranded.

Small businesses, too, found Subaru an invaluable partner. Shop owners used Sambar vans for deliveries, while construction workers and carpenters relied on the Leone for transporting tools and materials to remote job sites. Local governments and postal services adopted Subaru vehicles for use on secondary and tertiary roads, citing the brand's ability to tackle difficult terrain without excessive wear. In more remote mountain hamlets, a Sambar or Leone was often the only motor vehicle a family owned, used for everything from school runs to transporting livestock.

Subaru's engineering ethos was shaped by these rural realities. Every new model was subjected to extensive real-world testing on the sorts of roads its customers actually drove—unpaved, rutted, slick with snow or rain. The result was a line of vehicles with robust suspensions, sealed drivetrains, and engines designed to shrug off dust, mud, and hard use. Subaru's technical innovations—whether all-wheel drive, the low center of gravity from its boxer engines, or corrosion-resistant materials—were not marketing gimmicks but direct responses to the demands of rural life.

The impact of Subaru's rural roots extended to the company's internal culture. Engineers and product planners regularly visited rural regions to gather feedback and understand customer needs first-hand. Suggestions from farmers, shopkeepers, and mechanics shaped future product development, from easier-to-clean interiors to more practical loading areas and enhanced heater performance. This culture of listening and adapting fostered a cycle of improvement and loyalty, distinguishing Subaru from larger, less nimble competitors.

Even as economic growth and infrastructure improvements drew more Japanese to the cities, Subaru's presence in the countryside remained strong. The arrival of larger models like the Legacy and Outback in later decades did little to diminish the brand's rural following. On the contrary, these vehicles offered rural families the space, comfort, and all-weather capability required for longer commutes and growing aspirations, while maintaining the dependability that had built Subaru's name.

Rural Japan also played a vital role in Subaru's global story. Lessons learned on snowy mountain roads and muddy farm tracks informed export models destined for places with similar conditions—rural Australia, New Zealand, Scandinavia, the American Rockies, and the Canadian North. In each of these markets, Subaru's reputation for ruggedness and reliability traced its lineage directly to the fields and villages of Japan.

By the turn of the millennium, the sight of a Subaru—mud-splattered, loaded with produce or children or supplies—remained common in the Japanese countryside. Generations of families passed down stories of their first Subaru, of journeys through storms and winters, of a steadfast companion that never failed them. The loyalty born in these rural regions proved unshakable, shaping the brand's identity far beyond the boundaries of any city or country.

Subaru's enduring popularity in rural Japan stands as a testament to the power of meeting real needs with honest engineering and genuine care. It is a legacy built not on glamour or speed, but on the quiet confidence that comes from knowing a vehicle will start every morning, climb every hill, and carry its occupants safely home, no matter how rough the road or fierce the storm. This foundation, rooted deep in the countryside, would support Subaru as it looked outward to the world—carrying with it the lessons, loyalty, and spirit of rural Japan.

Chapter 7: The Leone Era: Expansion and Export

The dawn of the 1970s marked a pivotal moment for Subaru, as the company moved decisively from its early years as a domestic specialist to an automaker with international ambitions. Central to this transformation was the introduction and evolution of the Subaru Leone—a model whose name would become synonymous with Subaru's first true expansion beyond the borders of Japan. The Leone not only built on the company's growing reputation for dependability and technical ingenuity but also served as the vanguard for Subaru's global journey, most notably in North America and Australia, where the brand would establish deep roots.

The story of the Leone began in 1971, a time when Japanese car manufacturers were rapidly evolving, spurred on by increasing demand at home and rising curiosity abroad. The Leone was developed as a compact family car, positioned above the kei cars that had made Subaru's name in postwar Japan but below the larger sedans and wagons dominating international markets. The goal was to create a model line versatile enough to serve urban families, rural businesses, and, crucially, to appeal to a broad spectrum of overseas buyers with varying tastes and needs.

The first-generation Subaru Leone, introduced as both a two-door coupe and a four-door sedan, quickly distinguished itself through its engineering choices and styling. Its clean lines and understated design reflected the sensibilities of the era, while its compact dimensions, weighing just under a ton and stretching to just over four meters in length, made it ideal for both narrow Japanese streets and the growing demand for efficient cars in export markets. The most notable feature, however, was its optional four-wheel-drive system—a world first for a mass-produced passenger car. This innovation, originally conceived for the Japanese market's snow-prone regions, would prove pivotal in Subaru's global appeal.

Under the hood, the Leone continued Subaru's commitment to the horizontally opposed (boxer) engine layout. The initial 1.2-liter and 1.4-liter flat-four engines provided smooth power delivery and a low center of gravity, while their robust construction ensured reliability under a variety of driving conditions. As emissions standards tightened and performance expectations rose, subsequent iterations of the Leone would receive larger displacement engines, improved carburetion, and, eventually, fuel injection, reflecting the rapid technical progress of the 1970s and 1980s.

The introduction of the four-wheel-drive wagon variant in 1972 marked a watershed moment—not only for Subaru but for the global automotive industry. By integrating the practicality of a station wagon with the traction and capability of all-wheel drive, Subaru tapped into a unique niche. The Leone 4WD Wagon immediately found favor with customers living in mountainous and rural regions, as well as those facing regular bouts of snow and rain. The combination of comfort, efficiency, and go-anywhere ability resonated far beyond Japan's borders, catching the attention of importers and car buyers in markets with similar demands.

Subaru's move into North America was initially cautious but quickly gained momentum. In 1974, Subaru of America introduced the Leone as the DL/GL series, adapting the car to local preferences and regulations. The U.S. market, then dominated by large sedans and the first waves of imported compacts, was ripe for disruption. The timing could not have been better; the oil crisis of 1973 had sent fuel prices soaring, and American drivers were looking for alternatives to the thirsty V8s of Detroit. The Leone's efficient engines, simple mechanical layout, and the emerging popularity of the 4WD wagon struck a chord, especially in regions like New England, the Pacific Northwest, and the Rocky Mountain states, where challenging weather and terrain made all-wheel drive more than a luxury.

Australia, with its vast distances, punishing roads, and widely dispersed rural communities, was another market where the Leone found an enthusiastic reception. Local distributors quickly realized that the Leone's ruggedness, economy, and versatility filled a gap left by the decline of traditional rear-wheel-drive wagons and utes. Subaru's Australian operations invested in dealer networks and marketing campaigns that emphasized the Leone's ability to tackle everything from urban commutes to cross-country adventures. By the late 1970s, the Leone 4WD wagon had become a common sight on Australian farms, in mining towns, and even in motorsport competitions such as rallying, where Subaru's mechanical resilience and handling began to turn heads.

The Leone's evolution through its successive generations reflected not only the rapid changes in global automotive tastes but also Subaru's commitment to continuous improvement. The second generation, launched in 1979, brought a more angular, modern aesthetic, along with significant mechanical refinements. Engines grew in both power and sophistication, transmissions evolved from four-speed manuals to include five-speed options and, later, automatic gearboxes. Power steering, air conditioning, and more advanced safety features were added, keeping the Leone competitive against a rising tide of domestic and international rivals.

Subaru also responded to growing environmental concerns by developing cleaner engines and, in some markets, introducing front-wheel-drive versions for improved efficiency. Throughout the 1980s, the Leone's flexibility became one of its greatest assets. Buyers could choose from sedans, coupes, hatchbacks, and, most importantly, wagons—each offering Subaru's trademark reliability and, for those who needed it, the security of all-wheel drive.

The third generation, introduced in 1984, brought further advances in refinement, comfort, and technology. Fuel injection became increasingly common, turbocharged engines joined the lineup, and interior appointments approached levels expected from much larger and more expensive cars. The Leone continued to serve as Subaru's international workhorse, balancing innovation with the unpretentious practicality that had always defined the brand.

Export success was accompanied by a growing presence in motorsport, most notably rallying, where the Leone became the platform for Subaru's early international competition efforts. The car's all-wheel-drive system and balanced handling made it a natural contender in events characterized by rough roads, snow, and gravel. Victories in regional and national rally championships in Australia, New Zealand, and the United States burnished Subaru's reputation for ruggedness and technical prowess, setting the stage for the brand's later dominance with the Impreza.

In the U.S., Subaru's strategic advertising campaigns helped to solidify the Leone's image as the car for those who needed to "get there, no matter what." Commercials featured Leones fording streams, climbing snowbound hills, and traversing rural backroads, appealing directly to drivers in harsh climates or those seeking adventure. Subaru dealers became known for their straightforward, no-nonsense approach, attracting a customer base that valued substance over flash.

By the late 1980s, the market landscape had shifted. Consumer preferences in North America and Australia were evolving toward larger, more luxurious vehicles and the first waves of SUVs. Subaru responded by developing new models and laying the groundwork for the next era, but the Leone remained a steady seller and a brand cornerstone until its eventual replacement by the Legacy and Impreza platforms at the dawn of the 1990s.

Across its nearly two-decade lifespan, the Subaru Leone achieved production numbers exceeding 2.5 million units worldwide—a testament to its broad appeal, adaptability, and the trust it inspired. More than any single model before or since, the Leone was the bridge that carried Subaru from the narrow lanes and rural roads of Japan onto the world stage. It proved that a car could be both innovative and practical, affordable and rugged, and, most importantly, that Subaru could compete and thrive in the demanding international arena.

The legacy of the Leone endures, not only in the millions of cars that served faithfully across continents but also in the engineering principles and spirit of adventure that continue to define Subaru. Every Outback, Forester, and WRX carries forward the DNA of the Leone era—a period when a compact wagon with a boxer engine and a pioneering all-wheel-drive system showed the world that the road, no matter how rough or distant, was always within reach.

Chapter 8: Crossing the Pacific: Subaru's Entry into America

When Subaru set its sights on the United States in the late 1960s, the landscape of American motoring was at a crossroads. Detroit's "Big Three" still commanded the roads with their wide, V8-powered sedans, trucks, and muscle cars, but the ground was shifting. Economic uncertainty, mounting concerns over fuel costs, and growing interest in compact imports were creating new opportunities for automakers from overseas. Into this evolving market, Subaru entered not as a juggernaut, but as a thoughtful upstart, determined to find its place among giants and chart a different path.

Subaru's entry into America was preceded by years of cautious planning. Fuji Heavy Industries understood that simply shipping cars across the Pacific would not guarantee success. The United States was both vast and complex, with regional tastes, strict safety and emissions standards, and a population that, while increasingly receptive to imports, had developed strong preferences for certain brands and styles. Volkswagen's Beetle, the original Datsuns and Toyotas, and a handful of European makes had demonstrated that the American market could be won—but only with patience, adaptation, and an understanding of local needs.

In 1968, Subaru of America, Inc. was established in Bala Cynwyd, Pennsylvania. The company was founded in partnership with entrepreneur Malcolm Bricklin and Harvey Lamm, men with a keen eye for opportunity and a willingness to challenge convention. Subaru of America began modestly, with a small staff and a tiny inventory, aiming to introduce the Subaru 360 microcar to American drivers. The choice was bold: the 360 was ultra-compact, lightweight, and uniquely designed, its two-cylinder, 356cc engine producing only 25 horsepower. Bricklin, undeterred, hoped to market the car as "cheap and cheerful," offering mobility at a fraction of the cost of any domestic or imported rival.

Yet, the American debut of the Subaru 360 proved to be a trial by fire. While its charm and affordability attracted curiosity, the 360 was simply too small and underpowered for the vast highways and demanding driving conditions of the United States. Reviews in American automotive magazines were often unkind; Consumer Reports famously dubbed it "the most unsafe car in America," a verdict that, fair or not, hobbled the model's reputation from the outset. Sales were meager, and many early 360s languished on dealer lots.

Despite the inauspicious start, Subaru of America's leadership recognized both the challenge and the opportunity before them. The key, they realized, would be to move beyond the 360 and introduce models better suited to American preferences. The timing proved fortuitous. As the 1970s dawned, Americans were beginning to experience the shock of rising fuel prices and a changing regulatory climate. The Clean Air Act of 1970 and a wave of safety regulations forced all manufacturers to adapt, but they also created an opening for compact, efficient vehicles.

Subaru's answer was the FF-1 Star and, more consequentially, the Leone—introduced in the U.S. as the DL/GL series. These vehicles represented a dramatic evolution from the 360, offering more space, improved power, and modern amenities. The standout feature was the availability of four-wheel drive, first in wagon form and then across the lineup. Subaru became the first company to offer a mass-produced, affordable passenger car with four-wheel drive in America. In regions where winters were harsh and roads unpredictable—New England, the Rocky Mountains, and the Pacific Northwest—the 4WD Subaru quickly found a following.

Dealer development became a central priority. Subaru of America's leaders understood that establishing a credible, reliable dealer network would be critical to long-term success. Rather than pursuing rapid, unsustainable expansion, they focused on building relationships with independent dealers willing to take a chance on an unconventional brand. Training programs, parts supply improvements, and a focus on customer service helped to create the foundation for a network that prized loyalty and service over flashy showrooms.

Marketing campaigns were crafted to reflect Subaru's unique strengths and appeal to the pragmatic American buyer. Early ads did not pretend Subaru could compete with Detroit's power or luxury; instead, they leaned into themes of value, reliability, and ruggedness. "Inexpensive, and built to stay that way," became one of the brand's signature slogans. Print and television spots depicted Subarus tackling snowdrifts, muddy driveways, and mountain passes, connecting directly with the needs of rural families, outdoor enthusiasts, and drivers who valued getting through tough conditions over style or status.

The Subaru Brat (Bi-drive Recreational All-terrain Transporter), introduced in 1977, perfectly captured this spirit of functional quirkiness. Born from both American demand for small trucks and a creative response to the so-called "chicken tax" tariff on imported pickups, the Brat featured rear-facing jump seats in the cargo bed to qualify as a passenger vehicle. The Brat's unmistakable silhouette, go-anywhere attitude, and affordable price made it a cult favorite among young drivers, off-roaders, and those looking for something decidedly different from the ordinary pickup.

As the decade wore on, Subaru's reputation for dependability and all-weather performance continued to grow. Owners who bought their first Subaru out of necessity often became lifelong loyalists, impressed by the vehicles' ability to endure harsh conditions with minimal fuss. Mechanics appreciated the simple, robust design of Subaru engines and drivetrains, while families valued the peace of mind that came with all-wheel drive and a no-nonsense approach to motoring.

Subaru's image as an "outsider" in the American car market—neither Detroit iron nor mainstream Japanese import—was both a challenge and an asset. The company leaned into its outsider status, cultivating a sense of community and individuality among its owners. Subaru clubs, enthusiast groups, and grassroots motorsport events began to appear in snowy states and mountain regions, laying the groundwork for the passionate Subaru fandom that would flourish in later decades.

The 1980s brought continued evolution and growth. Models like the GL Wagon, with its optional dual-range transfer case, cemented Subaru's reputation among skiers, rural professionals, and drivers in the northern states. Subaru's advertisements, often lighthearted and self-aware, reinforced the brand's message: if you needed a car to get you through the toughest days—across snow-packed streets or up winding mountain passes—Subaru was your answer.

As Subaru's American footprint expanded, the company remained careful to never lose sight of its core values. It resisted the temptation to compete head-on with luxury brands or to dilute its identity in pursuit of volume. Instead, it focused on continuous improvement: refining all-wheel-drive technology, introducing fuel injection, improving safety, and developing models like the Justy, XT, and eventually the Legacy and Outback, each of which offered fresh takes on Subaru's formula of practical innovation.

By the end of the 1980s, Subaru had secured a distinct position in the crowded field of imports. While never the best-selling Japanese brand, it had become synonymous with all-weather confidence, durability, and a unique, sometimes eccentric approach to automotive design. Subaru's American success was hard-won and measured not in massive market share but in the loyalty of its customers and the affection of those who valued utility over fashion.

Crossing the Pacific had required vision, resilience, and a willingness to adapt without sacrificing identity. Subaru's story in America is a testament to the power of staying true to core principles—innovation rooted in real-world need, reliability in the face of adversity, and a quiet confidence that, no matter how tough the road ahead, there was always a way forward. For Subaru, the journey from a curiosity to a beloved fixture on American roads was just beginning, with greater challenges and triumphs still to come.

Chapter 9: The Brat and Beyond: Quirky Icons of the 1970s

As the 1970s unfolded, the global automotive landscape was being redrawn by changing regulations, shifting tastes, and a new generation of car buyers looking for alternatives to the mainstream. It was a decade marked by experimentation and upheaval, as oil crises and tightening emissions standards pushed manufacturers to think beyond the established template. For Subaru, a company already carving its own path with all-wheel-drive wagons and compact boxer engines, this was fertile ground for invention. The period gave rise to a series of offbeat, distinctive vehicles—most notably the Subaru BRAT—that would not only define the brand's personality, but also attract a new wave of buyers seeking something outside the norm.

The Subaru BRAT (Bi-drive Recreational All-terrain Transporter) arrived in 1977 as a direct response to unique circumstances on both sides of the Pacific. In America, the "chicken tax"—a 25% tariff on imported light trucks—posed a significant barrier to foreign automakers. Determined to offer a practical, affordable utility vehicle without incurring the steep penalty, Subaru engineers devised a now-iconic solution: they bolted a pair of rear-facing jump seats into the cargo bed, classifying the BRAT as a passenger vehicle and bypassing the import tax. The result was a small, nimble pickup with room for four and a distinctly unconventional appearance.

The BRAT's compact size, all-wheel-drive system, and lightweight construction made it a natural choice for buyers in rural, mountainous, or snow-prone areas. It was equally at home on farms, forest trails, or suburban driveways—ready to haul gear, tools, or friends to work and adventure alike.

The model's quirky features, such as the "jump seats," chrome roll bar, and side steps, turned heads wherever it appeared. Marketing materials leaned into this eccentricity, often showing the BRAT splashing through mud, scaling rocky inclines, or carrying surfboards to the beach. The car quickly gained a cult following among outdoor enthusiasts, young drivers, and anyone who valued the freedom to go places bigger vehicles simply could not.

The BRAT was powered by Subaru's proven flat-four engines, initially in 1.6-liter guise and later with larger displacements and optional turbocharging. The combination of modest weight and spirited power made for surprisingly lively performance, particularly off-road. The selectable 4WD system provided confident traction in slippery or uneven conditions, and the vehicle's simple, rugged construction endeared it to DIY mechanics and weekend tinkerers.

The appeal of the BRAT lay as much in its attitude as its abilities. At a time when most small trucks looked utilitarian and felt uninspired, the BRAT embraced its outsider status. It was, in many ways, a symbol of Subaru's willingness to challenge convention—not for its own sake, but to meet real-world needs with ingenuity and a dash of fun. The model's popularity in North America was mirrored in markets like Australia, where it was sold as the Subaru Brumby, and in New Zealand, where the "ute" tradition ran deep. In each region, the BRAT and its siblings found favor with buyers who needed a tool but also wanted a toy.

Subaru's 1970s product range was punctuated by other distinctive offerings that reinforced the brand's quirky reputation. The Subaru GL/DL wagons and sedans, for example, stood apart from competitors thanks to their boxer engines, part-time or full-time 4WD, and unique styling cues.

The "Cyclops" eye center grille, frameless windows, and unusual color options lent a playful quality to cars that were otherwise reliable and practical. Subaru also experimented with features such as push-button activated 4WD and "Twin Headlight" designs, always searching for ways to differentiate their cars from the competition.

The company's willingness to take risks extended to its marketing. Subaru advertisements in the late 1970s and early 1980s often featured humor, irreverence, and a direct challenge to automotive conventions. Slogans like "Inexpensive, and built to stay that way" or "We're not trying to be something we're not" resonated with buyers tired of empty promises and chrome excess. The BRAT itself starred in television spots that played up its rebellious character—an image cemented by real-life owners who put the car through every imaginable trial, from snowdrifts to sand dunes.

Beyond the BRAT, Subaru offered a variety of compact coupes, hatchbacks, and wagons, each tailored to the changing demands of a global customer base. Models like the Subaru FF-1 Star and the second-generation Leone demonstrated a growing maturity in design and engineering, blending dependability with innovation. The brand's engineering team continued to refine the boxer engine and 4WD system, gradually improving performance, emissions, and efficiency with each new model year.

One of the defining qualities of Subaru's 1970s and early 1980s vehicles was their adaptability. In North America and Australia, buyers prized the company's cars for their ability to double as family transport and weekend adventure machines. The wagons, in particular, became favorites among skiers, campers, and outdoor workers, often serving for decades in some of the harshest environments. Subaru's all-wheel-drive system—then still a novelty in the passenger car market—became a point of pride for owners, who delighted in showing off their vehicles' ability to conquer terrain that left conventional sedans stranded.

Subaru's embrace of the offbeat extended to its international product strategy. The brand was among the first Japanese manufacturers to recognize and respond to the specific needs of export markets. In Australia, for instance, the Brumby ute was tailored with heavy-duty suspension, increased ground clearance, and robust cooling systems to survive the Outback's punishing conditions. In the United Kingdom and Scandinavia, wagons were equipped with advanced heating and defrosting features for snowy climates. These adaptations were not mere marketing exercises but genuine responses to feedback from local dealers and drivers.

Controversy occasionally followed Subaru's quirky models. The BRAT's jump seats, designed as a tax dodge, provided little protection in an accident and were banned in some U.S. states. Yet, this controversy only added to the model's mystique and "outlaw" appeal, turning used BRATs into collector's items and conversation pieces. Even as safety and emissions standards evolved, Subaru's engineers found ways to retain the brand's signature character while improving compliance and usability.

By the early 1980s, Subaru's offbeat image had become one of its greatest assets. While other Japanese automakers focused on volume and uniformity, Subaru thrived by offering cars that were practical, affordable, and just a little bit weird. The BRAT and its siblings helped the brand weather the challenges of a volatile era, maintaining steady sales and growing a base of fiercely loyal customers. Subaru had succeeded in broadening its appeal well beyond traditional buyers, attracting a community of drivers who valued individuality, ingenuity, and the freedom to chart their own course.

The legacy of the BRAT and the quirky icons of the 1970s endures in the Subaru story. They proved that there was space in the automotive world for personality and playfulness, and that innovation could come not just from grand gestures, but from the simple act of solving problems in creative ways. For Subaru, these years cemented a reputation for reliability and resilience—but also for fun, irreverence, and an unapologetic sense of adventure that would guide the brand into the future.

Chapter 10: Surviving the Storm: Economic Challenges and Strategic Shifts

By the early 1970s, the automotive world was accelerating toward uncharted territory. Oil prices, long stable and predictable, suddenly spiked as a result of the 1973 oil crisis, sending shockwaves through economies and forcing car buyers, manufacturers, and governments alike to rethink their assumptions about transportation. For Subaru—then still a modest player among global giants—the coming decades would be a crucible, testing the company's resilience, adaptability, and strategic vision. The storm of economic upheaval, regulatory shifts, and changing consumer expectations would demand both courage and ingenuity, ultimately shaping Subaru into the brand recognized and respected worldwide today.

The first oil crisis, triggered by the OPEC embargo in 1973, was a wake-up call for automakers everywhere. Gasoline prices in the United States more than doubled within months, while rationing and long lines at filling stations became common scenes. For manufacturers of large, fuel-thirsty cars, the implications were dire; suddenly, the market shifted almost overnight in favor of compact, efficient vehicles. Japanese brands like Toyota, Honda, and Nissan surged in popularity, having already established a foothold with their smaller, frugal models.

Subaru's timing, though fortuitous in some ways, was not without challenges. Its early exports to the United States had focused on compact sedans, wagons, and the quirky BRAT—all of which were better suited to the new economic reality than the big V8s of Detroit. Yet, compared to larger rivals, Subaru's dealer network, brand recognition, and production capacity were limited. To survive and thrive, the company needed to make rapid, sometimes difficult decisions.

One key advantage for Subaru was its heritage of technical efficiency. The use of lightweight construction, compact boxer engines, and a growing focus on fuel economy had already defined the brand's approach. When the crisis hit, Subaru doubled down on these strengths. The company accelerated development of its most frugal models, refining carburetors, ignition systems, and engine management to eke out every possible kilometer from each liter of fuel. The Subaru Leone, introduced in the early 1970s, benefited from these advances and quickly became a staple among budget-conscious families and rural drivers.

In Japan, the domestic market responded to the oil crisis with an intensified appetite for kei cars and small, efficient family vehicles. Subaru's continued commitment to the kei class—particularly with the Sambar van and Rex hatchback—allowed the company to maintain solid sales and relevance despite a rapidly shifting landscape. The smaller engines and lightweight bodies of these models made them naturally economical, helping Subaru retain its rural and urban customer base alike.

The 1979 second oil shock, triggered by the Iranian Revolution, reinforced the lessons of the previous decade. Global fuel prices soared once more, and inflation gripped economies from the United States to Europe and Japan. Subaru's response was to invest in engineering flexibility, seeking to maintain quality and dependability while introducing ever-more efficient drivetrains. The company developed front-wheel-drive versions of its Leone and other models, offering buyers an even more economical alternative to four-wheel drive without sacrificing Subaru's trademark reliability.

Yet, economic turbulence also demanded broader strategic realignments. In the 1980s, with the global economy oscillating between recession and growth, Subaru embarked on a series of organizational reforms and partnerships. Recognizing the growing complexities of international trade and rising competition, Subaru (Fuji Heavy Industries) sought to strengthen ties with established overseas distributors and expand its research and development operations. The company invested in new manufacturing technologies, introducing automation and just-in-time production methods to control costs and improve quality.

At the same time, Subaru began looking toward international collaboration. This included limited partnerships and technology-sharing agreements with other automakers, laying the groundwork for later, deeper alliances. These moves helped buffer the company from the volatility of global markets, while allowing it to access new technologies and expertise.

Marketing also played a crucial role in navigating the storm. Subaru's advertising in the 1970s and 1980s was refreshingly straightforward, emphasizing real-world durability, economy, and the all-weather reliability of its cars. Campaigns such as "Inexpensive, and built to stay that way" resonated with buyers feeling the pinch of inflation. Subaru's focus on honesty, practicality, and value-for-money built customer loyalty even when budgets were tight.

Subaru's ability to pivot quickly and maintain a clear identity was tested time and again. The company avoided the temptation to chase fleeting trends, instead concentrating on incremental improvements and long-term dependability. Engineers continued to refine the boxer engine, improving emissions control systems and fuel efficiency, while maintaining the robust construction that had made Subaru a favorite in rural regions and challenging climates.

Another important survival tactic was diversification of product offerings. In addition to the mainstay Leone, Subaru introduced the Justy, a subcompact hatchback aimed squarely at the economy market. With its small footprint, lightweight design, and available four-wheel drive, the Justy quickly found favor in both domestic and export markets. It was also among the first Japanese cars to offer a continuously variable transmission (CVT), marking a forward-thinking approach to driveline efficiency that would become a Subaru hallmark in the decades to come.

While the core of Subaru's identity remained rooted in practicality, the company also recognized the need to keep pace with changing consumer expectations for comfort and technology. The third generation Leone, introduced in the 1980s, offered more refined interiors, improved safety features, and new options such as power steering and air conditioning. These enhancements helped Subaru maintain appeal as buyers became more discerning even in a cost-conscious environment.

The company's resilience was further tested by the fluctuating value of the yen, trade tensions, and a crowded field of domestic competitors. While Toyota, Honda, and Nissan expanded rapidly, Subaru focused on maintaining profitability and protecting its reputation for quality. The emphasis on all-wheel drive, a still-unusual feature among mass-market cars, continued to differentiate Subaru from its rivals and attract a dedicated following among drivers facing challenging road conditions.

During these decades of turbulence, Subaru's corporate culture played a decisive role in its survival and growth. The company's engineers and executives remained close to their customer base, gathering feedback, visiting dealerships, and personally investigating warranty claims and breakdowns. This commitment to constant improvement and customer satisfaction created a virtuous cycle—each new model was better suited to real-world needs, strengthening the brand's image and broadening its appeal.

Internationally, Subaru's steady hand and clear product philosophy paid off. The company expanded its presence in Australia, New Zealand, and select European markets, capitalizing on its reputation for durability in demanding environments. In the United States, Subaru's all-wheel-drive wagons became the vehicle of choice for skiers, hikers, teachers, and rural professionals, especially in the Snow Belt and Rocky Mountain regions.

By the end of the 1980s, Subaru had survived the economic storms that sank or humbled many competitors. Through strategic realignment, disciplined engineering, and a focus on practical value, the company had not only weathered the crises but emerged stronger, more agile, and better positioned for growth. The lessons learned during these challenging years—about efficiency, customer focus, and the importance of staying true to one's strengths—would form the bedrock for Subaru's global successes in the decades to come.

In the face of volatility, Subaru's commitment to honest engineering and practical solutions ensured that, even when the world changed overnight, its cars would keep starting in the morning, keep crossing mountains and muddy roads, and keep earning the trust of drivers from Japan to America and beyond. It was not always the flashiest or the fastest path, but in surviving the storm, Subaru had quietly built the foundation for its future as an enduring icon of resilience, innovation, and dependability.

Chapter 11: Legacy of Innovation: The Subaru Legacy and Modernization

By the late 1980s, Subaru stood at a crossroads, forged by decades of survival in the face of economic turbulence and technical adversity, yet still viewed in many markets as a maker of quirky, practical, and often utilitarian vehicles. The company's reputation for durability, all-wheel-drive expertise, and technical distinctiveness had earned it loyal customers in rural Japan, the American Snow Belt, and the wilds of Australia. Still, as global expectations for comfort, refinement, and technology advanced, Subaru recognized that long-term growth and international respect demanded something more—a leap forward in both engineering and brand ambition. The answer would be a car named not for the past, but for a vision of progress: the Subaru Legacy.

The birth of the Legacy in 1989 was a decisive moment in Subaru's history. The project began years earlier, within Fuji Heavy Industries, as engineers and product planners wrestled with how to transform the company's image and future prospects. The brief was ambitious: design and build a car that could go head-to-head with the world's best midsize sedans—not only in reliability and utility, but also in performance, refinement, safety, and style. The Legacy would be Subaru's most technically advanced vehicle to date, intended to compete not just with the likes of Toyota and Nissan, but with established European and American brands as well.

The development process was a showcase of Subaru's growing confidence and capacity for innovation. Engineers drew on the company's core strengths horizontally opposed (boxer) engines and all-wheel-drive systems—while pushing the boundaries of quality, aerodynamics, and manufacturing efficiency. Every aspect of the car was scrutinized for improvement, from noise insulation and interior materials to chassis rigidity and safety engineering.

The Legacy debuted in Japan in February 1989, followed quickly by international launches in North America and Europe. Its exterior design, penned with global tastes in mind, was sleek and modern, characterized by smooth lines, flush-fitting glass, and a distinctive wedge-shaped silhouette. The Legacy was larger than previous Subarus, with a longer wheelbase and wider track that offered increased interior space and a more commanding road presence. Inside, buyers found a carefully designed cabin with improved ergonomics, upgraded materials, and features such as electronically adjustable seats, advanced climate control, and (in higher trims) premium audio systems.

The engineering beneath the surface reflected both Subaru's heritage and its modernization. The Legacy was powered by a new generation of boxer engines, available in 1.8- and 2.0-liter displacements in Japan and a 2.2-liter unit for North America, each tuned for smoothness, fuel efficiency, and low emissions. Buyers could choose between front-wheel drive or Subaru's ever-advancing all-wheel-drive system, which now featured a sophisticated center differential and, in some versions, electronic traction management.

The chassis was equally innovative. The Legacy rode on a newly developed platform that prioritized both strength and lightness, utilizing extensive computer-aided design to minimize weight without sacrificing rigidity. The suspension—a combination of MacPherson struts up front and independent rear wishbones—offered a blend of ride comfort and handling precision that moved Subaru into a new class of driving dynamics. Safety was enhanced with features like anti-lock brakes, improved crash structures, and, in later versions, airbags and traction control.

Marketing the Legacy was an exercise in rebranding. Subaru positioned the car as a symbol of progress, blending the reliability and practicality of its predecessors with a fresh emphasis on sophistication and global competitiveness. The Legacy name itself—suggesting both heritage and forward momentum—was chosen to resonate with buyers seeking a car that would stand the test of time. Advertising campaigns highlighted the Legacy's advanced engineering and all-weather capabilities, as well as its ability to meet the demands of discerning drivers. In the U.S., the slogan "What to Drive" positioned the Legacy as an intelligent choice for those who wanted more than just basic transportation.

One of the most visible and daring demonstrations of Subaru's new confidence came in 1989, when the company organized a world-record endurance run at the Arizona Test Center. Three production Legacy sedans ran non-stop for 18 days, covering over 100,000 kilometers at an average speed of 223.345 km/h (138.78 mph)—a feat that set a new benchmark for speed and reliability and was publicized globally. The accomplishment gave Subaru powerful proof of the Legacy's quality, earning the brand attention in markets where it had previously been overlooked.

The launch of the Legacy coincided with another strategic move: the establishment of Subaru's first overseas production facility, Subaru-Isuzu Automotive, Inc. (SIA), in Lafayette, Indiana. Opened in 1989, the plant underscored Subaru's commitment to the North American market, reducing currency risk, improving supply chains, and allowing the company to tailor products more closely to local preferences. The Indiana plant would become a pillar of Subaru's growth strategy and a symbol of the brand's international maturity.

As the 1990s progressed, the Legacy continued to evolve, adding wagon and Outback variants that would become global icons in their own right. The Legacy Touring Wagon, launched in 1989, offered a uniquely practical and stylish alternative to traditional sedans and became especially popular in Japan and Europe. In 1995, Subaru introduced the Legacy Outback in the U.S., a crossover-inspired wagon with increased ground clearance, rugged styling, and a focus on adventure—the forerunner to an entirely new vehicle segment.

Technical innovation remained at the heart of the Legacy story. Turbocharged variants, introduced in Japan and Europe, offered performance on par with established sports sedans, while Subaru's relentless improvement of its all-wheel-drive technology cemented its reputation for all-weather confidence. The introduction of Active Torque Split and VTD (Variable Torque Distribution) systems provided ever-more sophisticated management of power delivery, enhancing both handling and safety.

The Legacy also set new standards for environmental responsibility. The Indiana plant was recognized as one of the first zero-landfill automobile factories in the United States, reflecting Subaru's growing commitment to sustainability. The brand's focus on low-emissions powertrains and responsible manufacturing practices appealed to a new generation of environmentally conscious buyers.

The influence of the Legacy on Subaru's direction was both immediate and enduring. The model's success proved that Subaru could compete in the demanding global midsize market, establishing a template for future models in terms of quality, technology, and international appeal. The Outback, born as a Legacy variant, would become a best-seller and a symbol of Subaru's adventurous spirit, while the Legacy platform itself provided the foundation for subsequent innovations, including the development of the sporty Impreza and WRX.

Critically, the Legacy era transformed perceptions of Subaru from a niche player to a serious contender. Awards, positive reviews, and strong sales numbers confirmed that the company's investment in quality and modernization had paid off. Subaru was no longer seen solely as the maker of quirky 4WD wagons and microvans; it was now a brand associated with forward-thinking engineering, robust safety, and the ability to blend practicality with refinement.

For Subaru, the Legacy represented more than just a successful car launch—it was a statement of purpose, a demonstration of the company's willingness to learn from the past, embrace change, and aspire to new heights. The lessons of the Legacy would echo through every model that followed, guiding Subaru's ongoing pursuit of innovation and excellence. In redefining its ambitions for quality and global competitiveness, Subaru set a new course for the decades ahead, ensuring that the spirit of progress—true to the Legacy's name—would always remain at the heart of the brand.

Chapter 12: Rally Roots: Subaru's Ascent in World Motorsport

In the early 1990s, as Subaru's road cars reached new heights of refinement and global ambition with models like the Legacy and Impreza, another facet of the company's identity was coming into focus—one built not in factories or showrooms, but on gravel tracks, snowy mountain passes, and rain-lashed forest stages. This was the world of rallying, a motorsport where technical ingenuity, resilience, and adaptability mattered as much as outright speed. For Subaru, rallying would not only be a proving ground for its signature technologies, but also a global stage upon which the brand's reputation would be transformed from a specialist maker of rugged family cars into an icon of all-wheel-drive performance and adventure.

Subaru's involvement in rallying began modestly. In the late 1980s, the brand's all-wheel-drive wagons and sedans had already built a loyal following among enthusiasts, particularly in snowy or mountainous regions where rallying's spirit was part of everyday driving. Subaru's technical team recognized the parallels: the need for robust suspension, the advantages of balanced boxer engines, and—most critically—the capabilities of advanced all-wheel-drive systems. The debut of the Legacy in 1989 gave Subaru an ideal platform to enter international rally competition, a move that was as much about engineering development as marketing.

The creation of Subaru Tecnica International (STI) in 1988 marked a turning point. STI was established as the company's dedicated performance and motorsports arm, with a mandate to develop high-performance vehicles, coordinate motorsport efforts, and bring technological advances from the rally stages to the production line. The first major STI project was preparing the Subaru Legacy RS for the World Rally Championship (WRC), the sport's most demanding and prestigious series.

Subaru entered the WRC in 1990, fielding a team that combined technical expertise from Japan with British motorsport know-how through a partnership with Prodrive, a renowned motorsport engineering company based in Banbury, England. The Legacy RS, powered by a turbocharged, 2.0-liter boxer engine and equipped with a sophisticated all-wheel-drive system, was immediately competitive, though not yet a title contender. Drivers like Markku Alén, Ari Vatanen, and later, a young Scottish talent named Colin McRae, began to demonstrate the car's potential on rally stages from Sweden to Kenya.

The early years in the WRC were a period of learning and rapid evolution. The Legacy RS, though strong and reliable, was relatively large and heavy compared to some of its rivals. Subaru and Prodrive responded by focusing on incremental improvements: reducing weight, fine-tuning suspension setups, developing faster-responding all-wheel-drive systems, and extracting more power from the turbocharged engine. These relentless refinements yielded results, with podium finishes and stage wins confirming Subaru's growing competitiveness.

The real breakthrough, however, came with the introduction of the Impreza. Launched as a road car in 1992, the Impreza offered a shorter wheelbase, lighter body, and improved balance compared to the Legacy. In 1993, Subaru debuted the Impreza 555 rally car—a machine that would soon become legendary. With its powerful 2.0-liter turbo boxer engine, advanced symmetrical all-wheel-drive, and distinctive blue-and-yellow livery, the Impreza 555 combined Subaru's engineering DNA with a razor-sharp focus on motorsport performance.

The partnership with Prodrive and the arrival of charismatic drivers—most notably Colin McRae—propelled Subaru to the forefront of world rallying. McRae, renowned for his fearless driving style and determination, quickly became a fan favorite and the face of Subaru's WRC campaign. The 1995 season proved to be the watershed moment: McRae won the World Rally Championship Drivers' title, becoming the first Briton ever to do so, and Subaru clinched the Manufacturers' title, cementing its status as a rally powerhouse.

The technical story behind these victories was as compelling as the sporting drama. Subaru's rally cars became showcases for cutting-edge technologies: electronically controlled center differentials, active differentials for both axles, sophisticated turbocharging systems, and relentless fine-tuning of suspension geometry. The symmetrical layout of Subaru's all-wheel-drive system, paired with the boxer engine's low center of gravity, gave the cars remarkable stability and traction on loose, unpredictable surfaces. Whether on the ice roads of Sweden, the rocky tracks of Greece, or the rain-soaked forests of Wales, Subaru rally cars were celebrated for their ability to deliver both speed and control.

The rallying successes had an immediate and dramatic impact on Subaru's global reputation. The distinctive blue-and-gold Impreza became a hero car, capturing the imagination of enthusiasts and everyday drivers alike. Road-going versions—such as the Impreza WRX and WRX STI—became instant icons, coveted for their rally-bred technology and aggressive performance. Subaru's rally image helped the company attract a younger, more passionate customer base, and cemented all-wheel-drive as not just a practical feature, but an enthusiast's hallmark.

Beyond the World Rally Championship, Subaru's rally cars dominated national and regional competitions across the globe. In the United States, Canada, Australia, and New Zealand, Subaru vehicles became fixtures on the rally scene, racking up victories and building grassroots support. In Japan, Subaru's rally exploits were a source of national pride, reinforcing the brand's commitment to technical excellence and competitive spirit.

The 1990s and early 2000s saw Subaru and Prodrive continue to innovate, with successive generations of the Impreza WRC car incorporating advances in materials, aerodynamics, and electronics. The team's relentless pursuit of performance paid off with three consecutive WRC Manufacturers' titles from 1995 to 1997. Subaru's rally cars became a test bed for technologies that would later appear in production vehicles, from improved turbocharging and intercooling systems to stronger transmissions and advanced all-wheel-drive management.

Rallying also shaped Subaru's marketing and brand culture. Advertising campaigns highlighted the direct connection between the rally stages and the showroom floor. Phrases like "World Rally Champion" and "Rally Proven" appeared in brochures and commercials, while images of airborne Imprezas and mud-splattered Leones became central to Subaru's identity. Limited-edition models and special color schemes—most famously the WR Blue and gold wheels—celebrated the company's motorsport heritage.

For Subaru, rallying was never just about racing—it was about testing technology under the harshest conditions and proving the real-world benefits of engineering choices. The lessons learned from rally competition—about resilience, adaptability, and the pursuit of excellence—permeated every aspect of the company's operations, from research and development to customer service and brand messaging.

The legacy of Subaru's rally era endures not only in its trophy cabinet but in the DNA of every Subaru vehicle that followed. The brand's ascent in world motorsport transformed how Subaru was perceived: no longer simply a maker of practical, dependable cars for the countryside, Subaru became synonymous with adventure, innovation, and winning against the odds. The rally roots remain visible in the WRX and STI badges, in the global community of Subaru enthusiasts, and in the company's ongoing commitment to engineering distinction. Subaru's journey from gravel tracks to global acclaim stands as a testament to the power of competition, teamwork, and the relentless quest for improvement—principles as vital on the rally stage as they are on the road.

Chapter 13: Blue and Gold: The Impreza WRX Arrives

By the mid-1990s, Subaru's technical and sporting credibility had reached new heights thanks to a potent combination of rally success and relentless engineering. The company's ascent in the World Rally Championship had delivered not just trophies, but global recognition and a powerful new image—one inseparable from the thunderous sound of a turbocharged boxer engine, the unmistakable flash of blue paint streaked with gold, and a single, iconic name: Impreza WRX.

The origins of the Impreza WRX can be traced to the launch of the standard Impreza in 1992. Intended as a compact, agile, and affordable car for global markets, the Impreza inherited much of its engineering from the larger Legacy, including the option of Subaru's renowned all-wheel-drive system and, in performance variants, the horizontally opposed boxer engine. However, where the Legacy aimed for refinement and family practicality, the Impreza's lighter chassis and tighter proportions made it a natural candidate for motorsport development—and for a new breed of performance-oriented customer.

Within Subaru, the desire to translate rally experience into a production car was driven by both marketing acumen and technical ambition. The company's World Rally Championship program was beginning to deliver results, but there was recognition that a road car echoing the spirit—and even the specifications—of the rally machines could become a new halo model, raising the brand's profile and offering enthusiasts something truly unique. The result was the Impreza WRX—World Rally eXperimental—a car that would, in just a few short years, become a global phenomenon.

Launched initially for the Japanese market in late 1992, the first Impreza WRX was a revelation. At its heart was a turbocharged 2.0-liter DOHC boxer engine, producing upwards of 240 horsepower in top trim and matched to a close-ratio five-speed manual gearbox. Subaru's symmetrical all-wheel-drive system, already proven on rally stages, came standard, giving the WRX exceptional traction and poise in all conditions. The car's relatively light weight, rigid chassis, and sport-tuned suspension made for razor-sharp handling, while disc brakes at all four corners and aggressive gearing delivered stopping power and rapid acceleration.

It was not just the technical specification that set the WRX apart, but the raw immediacy of the driving experience. The turbocharged boxer thumped and burbled with a sound unlike any rival, and as the boost came in, the WRX launched forward with startling pace. The car's character—intense, playful, always eager—quickly earned it a devoted following in Japan, where street and circuit racing scenes were blossoming and young enthusiasts were hungry for new legends.

In parallel with the standard WRX, Subaru Tecnica International (STI)—the company's motorsport and performance division—developed even more focused variants. The Impreza WRX STI debuted in 1994, boasting further upgrades: more power, stiffer suspension, advanced differentials, and lighter components. Each STI was hand-assembled with care, drawing upon lessons learned in the heat of rally competition. The STI badge soon became a mark of exclusivity and capability, signifying a car as close as possible to Subaru's WRC machines while remaining road legal.

One of the defining visual motifs of the WRX and WRX STI era was the adoption of blue paint—specifically, Subaru's World Rally Blue Mica—and gold wheels. This now-legendary color scheme had its origins in the Prodrive rally team's livery, sponsored by 555 State Express cigarettes, which used deep blue and gold as their corporate colors. The combination became so closely associated with Subaru's performance image that even after the end of tobacco sponsorship, it remained the unofficial uniform of the WRX and STI lines. For a generation of enthusiasts, blue and gold became visual shorthand for rally pedigree, speed, and Subaru's outsider appeal.

As Subaru's rally victories mounted—most notably the manufacturers' titles in 1995, 1996, and 1997—the Impreza WRX's reputation grew far beyond Japan. Grey-market imports to the United Kingdom, Australia, and elsewhere quickly sold out, and the WRX began to acquire near-mythical status among enthusiasts who saw it in magazine road tests, television coverage of rally events, and popular culture. The car's rally-derived technology—turbocharged engines, all-wheel drive, multi-mode differentials, and advanced suspension geometry—was unmatched by mass-market rivals, and its underdog status only heightened its allure.

In 1994, Subaru began officially exporting the WRX to several markets outside Japan, with the United Kingdom among the first to receive it in right-hand-drive form. The car's impact on the British and Australian performance scenes was immediate and profound. The WRX rapidly became a dominant force in rallying at every level, from club events to national championships, while in everyday driving, it offered a combination of practicality and performance that few could rival. The car's signature blue and gold made it instantly recognizable on both road and stage.

In the United States, enthusiasts watched the WRX's exploits from afar for nearly a decade, as Subaru hesitated to bring the car stateside amid concerns over insurance, emissions, and the perceived demand for high-performance imports. That changed in 2001, when the second-generation "bug-eye" WRX finally reached American shores. The U.S.-spec WRX, while slightly detuned for emissions compliance, still offered a turbocharged boxer engine, all-wheel drive, and the visceral, unfiltered driving experience that had become the model's calling card. American car magazines heaped praise upon the WRX, hailing it as a game-changer for affordable performance, and sales exceeded all expectations.

With each subsequent generation, the WRX and WRX STI continued to evolve in both performance and refinement. Power increased, handling became ever sharper, and advanced features such as multi-mode driver-controlled center differentials and Brembo brakes raised the bar for capability. The car's motorsport credentials were further cemented as STI-tuned Imprezas claimed victories in everything from international rallies to the punishing Nürburgring 24 Hours. Special editions, limited-production models, and a thriving aftermarket scene allowed owners to personalize and enhance their WRXs, deepening the car's cult status.

Beyond raw numbers and trophies, the Impreza WRX and STI reshaped perceptions of what Subaru—and by extension, Japanese automakers—could achieve in the realm of performance. The WRX became the car of choice for a new generation of enthusiasts, motorsport aspirants, and anyone who wanted genuine rally-bred engineering without sacrificing daily usability. Its distinctive exhaust note, squat stance, and blue-and-gold livery were celebrated not just in magazines and showrooms, but in video games, movies, and across the digital landscape of internet car culture.

The WRX phenomenon also reverberated throughout Subaru's lineup, inspiring the creation of sporty variants across the range and reinforcing the brand's commitment to all-wheel-drive performance. Even as the market shifted towards SUVs and crossovers, the WRX remained a symbol of Subaru's adventurous spirit, technical courage, and refusal to blend into the background.

For Subaru, the story of the Impreza WRX and WRX STI is one of aspiration realized. What began as an effort to channel rally success into a road car became a global legend, a machine that united engineering innovation, motorsport glory, and a distinctive, irrepressible personality. The image of a blue WRX with gold wheels—dust swirling, turbo spooling, boxer thumping—remains an enduring emblem of what Subaru can achieve when it dares to lead with passion, technology, and a sense of play. It is a legacy built not just on speed, but on the thrill of possibility and the conviction that greatness can come from those willing to chart their own course.

Chapter 14: The Turbocharged Era: Power Meets Practicality

The story of Subaru's evolution into a performance and engineering icon cannot be told without the transformative impact of turbocharging. While the concept of forced induction was hardly new—having roots in aviation and racing from the early twentieth century—Subaru's unique application of turbocharging across its mainstream models in the late twentieth and early twenty-first centuries marked a turning point not only for the brand, but also for a generation of car enthusiasts who discovered that thrilling power and daily usability could finally coexist in one package.

The earliest whispers of Subaru's turbocharged ambitions began in the 1980s, a decade when automakers worldwide experimented with turbo technology to balance increasing demands for fuel efficiency, rising expectations for performance, and tighter emissions standards. For Subaru, the challenge was both technical and philosophical: how to stay true to its core values of reliability, practicality, and all-weather confidence while satisfying an emerging appetite for sportiness and speed.

The company's first foray came in 1983 with the launch of the Subaru Leone Turbo. This model—available as a sedan, wagon, and coupe—was powered by a turbocharged 1.8-liter EA82 flat-four engine producing up to 111 horsepower. Compared to the normally aspirated variants, the Leone Turbo offered a distinct jump in power, torque, and highway passing ability, all while maintaining the famed Subaru dependability. The turbo system used a compact draw-through carburetor setup, a sign of both the era's technology and Subaru's determination to push the envelope despite limited resources.

What set Subaru's early turbocharged offerings apart was not just the engine, but the holistic package: all-wheel drive became available in turbocharged Leones, blending extra power with unmatched traction. In an age when many competitors still reserved four-wheel drive for rugged trucks or off-roaders, Subaru quietly demonstrated that forced induction and real-world practicality could live side by side. The result was a car that, while understated, was equally at home on the daily commute or a snowy mountain road.

As the 1980s turned to the 1990s, Subaru continued to refine its turbocharging technology. With the introduction of the Legacy in 1989, the company launched a new generation of turbocharged flat-four engines. The Legacy RS, particularly in Japan and Europe, featured a 2.0-liter DOHC turbocharged engine generating up to 220 horsepower and paired with full-time all-wheel drive and a five-speed manual transmission. This combination gave the Legacy not just straight-line speed, but a level of control and composure in adverse conditions that set new standards for family sedans and wagons.

The Legacy Turbo was more than just a faster Subaru. It represented the brand's first earnest attempt to compete in the growing global market for performance-oriented, yet practical, vehicles. The car's success on the road was matched by its exploits in motorsport, where the Legacy Turbo served as Subaru's entry into the World Rally Championship, laying the foundation for the even more legendary Impreza WRX. For customers, the Legacy Turbo signaled a new era: they no longer had to choose between utility and excitement.

It was the Impreza WRX, however, that brought turbocharging to the heart of Subaru's public image.

Launched in 1992, the WRX featured a turbocharged 2.0-liter boxer engine that, depending on the market, delivered between 218 and 276 horsepower. The WRX's relentless acceleration, distinctive exhaust note, and rally-inspired handling set a new benchmark for compact performance cars. Everyday usability remained at the core—five seats, a real trunk, sensible ergonomics—making it the perfect daily driver for those who craved excitement without compromising on family or weather needs.

The WRX STI, developed by Subaru Tecnica International, raised the bar even further. Enhanced turbocharging, intercooling, and advanced engine management pushed output higher, while close-ratio transmissions, advanced all-wheel-drive systems, and beefed-up brakes ensured that the extra power could be safely and enjoyably deployed. The STI's reputation for reliability in the face of hard driving, combined with its real-world versatility, helped establish Subaru's turbocharged performance models as not only fast, but also sensible long-term investments.

As turbocharged Subarus grew in fame, so too did the breadth of their appeal. In Europe, Australia, and Japan, the WRX and STI became the weapon of choice for rally fans, club racers, and spirited commuters. In North America, the arrival of the WRX in 2001 and the STI in 2004 ignited a performance revolution, introducing thousands of drivers to the joys of turbocharged all-wheel drive. The turbocharged Forester and Legacy models broadened the appeal even further, offering high-output performance in wagon, crossover, and sedan bodies that blended adventure with everyday functionality.

Turbocharging's influence extended beyond outright performance. In the face of ever-stricter emissions standards and consumer demands for fuel economy, Subaru engineers used turbocharged downsized engines to deliver the best of both worlds. By employing precise electronic fuel injection, variable valve timing, and advanced intercooling, Subaru created engines that could run smoothly in city traffic and surge powerfully on open roads—without the fuel penalties of larger, naturally aspirated engines. The brand's willingness to invest in the robust cooling and lubrication systems necessary for long-term turbo reliability further enhanced its reputation among buyers wary of temperamental performance cars.

Marketing and public perception shifted with this technological advance. Subaru's turbocharged lineup became central to the brand's identity: advertisements highlighted the excitement of turbo power, the security of all-wheel drive, and the everyday utility that came standard with every model. The image of a blue WRX cresting a snowbank, or a turbocharged Forester climbing a mountain trail, resonated deeply with a customer base that valued adventure as much as acceleration. The once-niche idea of combining speed and practicality was now mainstream, and Subaru was leading the charge.

Enthusiasts responded with fervor. The WRX and STI in particular inspired a vibrant tuning culture, with owners modifying their cars for more power, sharper handling, or unique looks. Aftermarket companies flourished, track days and rally events filled with Subarus became commonplace, and a global community of drivers emerged around the shared experience of turbocharged boxer power.

Critically, turbocharging did not remain exclusive to performance flagships. Subaru recognized that the benefits of forced induction—higher torque at low RPM, better response in all conditions, and the ability to tailor power delivery to different markets—could serve mainstream models as well. The Forester XT, Legacy GT, and Outback XT brought turbo power to families, adventurers, and commuters alike, each model designed to deliver a blend of everyday comfort and effortless acceleration.

In time, Subaru's turbocharged vehicles became a symbol of balance: excitement without excess, capability without compromise, and innovation grounded in real-world needs. While the era of large, naturally aspirated engines waned and hybrid technologies gained ground, the enduring appeal of Subaru's turbocharged models stood as proof that power and practicality, when engineered with care, could coexist.

The turbocharged era was not just a chapter in Subaru's technical evolution, but a defining philosophy. It represented the company's belief that a car could—and should—do more than simply get its occupants from A to B. Subaru proved that the journey could be thrilling, that performance need not come at the expense of reliability or comfort, and that a carmaker committed to innovation could change what drivers expected from every trip. Power, practicality, and poise: in the turbocharged era, Subaru made all three the standard, not the exception.

Chapter 15: Outback Bound: Pioneering the Crossover Movement

In the mid-1990s, as the sport utility vehicle boom began reshaping automotive tastes around the world, Subaru faced a crucial crossroads. The company's reputation for practical wagons, dependable all-wheel-drive technology, and understated engineering was well established among loyal customers, but the marketplace was changing. Shoppers increasingly wanted vehicles that combined the comfort and drivability of a passenger car with the rugged looks and capabilities of traditional SUVs. Most automakers responded by launching new, truck-based models or developing larger, thirstier vehicles—an approach that risked alienating buyers who valued efficiency and everyday utility. Subaru, with characteristic resourcefulness, took a different path. The result was the Subaru Outback: an unassuming yet revolutionary blend of wagon and SUV that not only created a new automotive segment but also propelled Subaru into the center of a global shift toward crossovers.

The origins of the Outback can be traced to Subaru of America in the early 1990s. The U.S. market was undergoing rapid change as buyers abandoned conventional station wagons in favor of the higher ride heights and commanding presence of SUVs. Subaru's sales, though steady, faced mounting pressure from larger, flashier, and more heavily marketed competitors. Executives and dealers recognized a challenge: how could Subaru capitalize on the SUV craze without the expense, risk, and complexity of developing a new, truck-based platform from scratch?

Subaru's answer was characteristically inventive. Instead of starting anew, the company took the successful Legacy wagon—already renowned for its symmetrical all-wheel-drive system, practicality, and reliability—and reimagined it for a new generation. Engineers and designers raised the ride height, added tougher cladding and bumpers, equipped it with larger wheels and tires, and included functional roof racks and splash guards. The mechanical heart remained unchanged: a boxer engine under the hood and a full-time all-wheel-drive system beneath, offering the sure-footedness of an SUV without the weight or unwieldiness.

The name "Outback" was carefully chosen, evoking the spirit of adventure, resilience, and vast open spaces—qualities that resonated with outdoors enthusiasts and suburban families alike. Subaru's marketing team leaned heavily into this narrative, launching the new model in 1994 for the Japanese market and 1995 for North America. American advertisements memorably starred Paul Hogan, the actor best known for playing Crocodile Dundee, whose rugged charm and Australian accent quickly became synonymous with the Outback's personality.

The first-generation Outback was an immediate success. Buyers responded to its promise of go-anywhere ability paired with everyday sensibility. The Outback's car-based platform delivered a smooth, composed ride and better fuel economy than most truck-based SUVs. At the same time, its increased ground clearance and robust all-wheel-drive made it credible in snow, mud, and gravel. Subaru's gamble on the crossover concept was vindicated almost overnight, as sales surged in regions with harsh weather, active lifestyles, and a growing desire for vehicles that looked tough but handled city streets with ease.

The Outback's practical advantages set it apart from both traditional wagons and emerging SUVs. Its long roof and rear hatch offered abundant cargo space, while folding rear seats made it easy to haul sports equipment, pets, and home improvement supplies. Families found the Outback's manageable size perfect for urban errands and school runs, while adventurers appreciated its ability to tackle mountain passes and forest trails. With a low step-in height and wide-opening doors, the Outback remained accessible for children, seniors, and those who simply preferred ease of use over the climbing and clambering demanded by taller vehicles.

Subaru's engineers continued to refine the Outback formula with each new generation. The second-generation Outback, launched in 1999, introduced a more powerful engine lineup, improved crash safety, and even higher ground clearance. Special attention was paid to durability, with underbody protection, corrosion-resistant materials, and robust suspension components engineered for real-world abuse. The model range soon expanded to include six-cylinder engines, turbocharged variants, and luxury trims, ensuring the Outback's appeal spanned from frugal commuters to premium buyers.

Internationally, the Outback found receptive audiences wherever rugged terrain and challenging weather were part of daily life. In Australia—the source of the Outback's name and marketing mythology—the model earned a reputation for outlasting the competition in the bush and on the highway. In Europe, Scandinavia, and Canada, the Outback became the car of choice for those who needed secure winter traction and generous cargo space without sacrificing comfort or economy. The Outback's versatility made it popular with government agencies, mountain rescue teams, veterinarians, and anyone whose life or work required dependable, go-anywhere transport.

The Outback's runaway success did more than bolster Subaru's sales figures; it fundamentally redefined the brand's image in key markets. Subaru, long known for quirky small cars and rural dependability, was suddenly on the cutting edge of automotive trends. The Outback became a cultural touchstone, celebrated in everything from outdoor magazines to television shows and national advertising campaigns. Its blend of utility, safety, and approachable personality attracted new customers, particularly in North America, where it became a mainstay of active suburban families, outdoor adventurers, and even urban professionals drawn by its practicality and understated cool.

The Outback's influence extended well beyond Subaru's own lineup. As sales soared and rivals took notice, the entire industry began to pivot toward the crossover segment. Automakers large and small launched a wave of vehicles that adopted the Outback's formula: car-based chassis, available all-wheel drive, versatile interiors, and a rugged appearance. The term "crossover" entered the automotive lexicon, and Subaru's foresight in bridging the gap between wagons and SUVs was recognized as a key catalyst for this market shift.

Technically, the Outback continued to innovate. Successive generations brought advances such as continuously variable transmissions (CVTs), direct-injection turbocharged engines, advanced safety systems (including Subaru's EyeSight driver assistance), and hybrid technology. All the while, the Outback retained its essential qualities: the low center of gravity of a boxer engine, the confidence of symmetrical all-wheel drive, and the honesty of a design that prioritized function as much as form.

The Outback also became a platform for Subaru's commitment to safety and environmental stewardship. Subaru invested heavily in crash protection, earning top scores from global safety agencies year after year. Its manufacturing plants, including the Lafayette, Indiana facility where Outbacks for North America are built, became industry leaders in zero-landfill, sustainable operations. Subaru's sponsorship of environmental causes and national parks further cemented the Outback's association with outdoor adventure and responsible stewardship.

For Subaru, the Outback's success was not just commercial, but transformational. It allowed the company to weather periods of intense competition, economic downturns, and shifting consumer preferences while maintaining a distinct identity. The Outback's loyal customer base—often among the most devoted in the automotive world—became ambassadors for the brand, spreading its virtues by word of mouth, online forums, and social media.

In the broader history of the automobile, the Subaru Outback stands as a genuine pioneer. It proved that ruggedness and refinement, adventure and everyday comfort, were not mutually exclusive. By daring to blend the practical virtues of a wagon with the tough spirit of an SUV, Subaru didn't just find a new path for itself—it carved out a road that the entire industry would soon follow. The Outback's legacy endures in millions of driveways, trailheads, and campsites worldwide, a testament to the enduring appeal of a vehicle born from innovation, necessity, and the courage to do things differently.

Chapter 16: Keeping It Simple: Kei Cars and Urban Mobility

Amid the dramatic advances in turbocharging, all-wheel drive, and performance engineering that made Subaru a global icon, another equally important chapter in the brand's history unfolded in the bustling streets and narrow lanes of Japan's cities. Here, the focus was not on rally victories or cross-country adventures, but on practicality, innovation, and the everyday realities of urban life. For decades, Subaru's commitment to the kei car segment—a uniquely Japanese class of small, efficient vehicles designed for maximum utility in tight spaces—demonstrated the company's versatility, creativity, and enduring connection to the needs of ordinary drivers. Through models like the R-2, Rex, and Vivio, Subaru played a formative role in shaping how Japan moved, worked, and lived, embodying the philosophy that true engineering excellence is measured as much by simplicity as by sophistication.

The concept of the kei car (keijidōsha, or "light automobile") originated in postwar Japan, when the government introduced a set of regulations that capped engine displacement, vehicle size, and power output to encourage affordable mobility, reduce urban congestion, and conserve resources. Kei cars benefited from significant tax, insurance, and parking incentives, making them an appealing option for city dwellers and rural families alike. For automakers, the segment presented a challenge: to maximize utility, comfort, and style within the strictest of technical constraints.

Subaru's involvement with kei cars began in earnest in the late 1960s, following the success of the 360 microcar and the Sambar light commercial vehicles. In 1969, the company launched the Subaru R-2, a successor to the 360 designed from the ground up to address evolving tastes and regulatory requirements. The R-2 retained the rear-engine, rear-wheel-drive layout of its predecessor but featured a larger, more modern body with improved interior space, better visibility, and a smoother ride.

Under the hood, the R-2 initially relied on an air-cooled, two-stroke engine—a technology that had served Subaru well for years. However, as emissions standards tightened and consumer expectations shifted, Subaru began developing water-cooled four-stroke units for future models. The R-2's compact footprint, nimble handling, and accessible price point quickly won over a new generation of urban drivers. Its popularity helped Subaru further entrench itself in the competitive kei car market, competing with giants like Honda's N360, Suzuki's Fronte, and Daihatsu's Fellow.

The evolution of Subaru's kei cars mirrored changes in Japanese society itself. Rapid urbanization during the 1970s created dense, vertical cities where every meter of road and parking space was at a premium. The R-2's successors would respond directly to these pressures. In 1972, Subaru introduced the Rex—a name derived from the Latin word for "king," reflecting the company's aspirations for the model's leadership in the segment. The first-generation Rex retained the essential simplicity and compactness of its forerunners but incorporated a front-engine, rear-wheel-drive layout for better weight distribution and improved safety.

As regulations continued to evolve, Subaru showed remarkable agility in adapting its kei car lineup. The second-generation Rex, launched in 1976, featured an all-new, water-cooled four-stroke engine—a significant step forward for performance, emissions, and refinement. The engine, known as the EK series, was available in both naturally aspirated and supercharged forms, offering brisk acceleration while remaining within the segment's strict power limits. The Rex's tall, squared-off body maximized interior space, while innovations such as fold-flat seats, sliding doors, and configurable cargo areas set new standards for urban utility.

Subaru's kei cars were notable not just for their practicality but also for their quiet ingenuity. The company's engineers continually sought new ways to improve ride comfort, noise insulation, and ease of maintenance. Features such as fully independent suspension, rack-and-pinion steering, and high-efficiency heating and ventilation systems made the Rex and its siblings unusually comfortable for their size. Safety was another focus: reinforced passenger cells, improved crumple zones, and advanced braking systems offered reassurance to buyers navigating the hazards of crowded city streets.

In the 1980s, the kei car segment became even more competitive and technologically advanced. Subaru responded with the Rex Combi, a five-door hatchback variant that offered greater versatility for families and small businesses. The company also introduced supercharged versions of the Rex, using compact forced induction to squeeze maximum performance out of the limited displacement engines. These models delivered zippy acceleration, lively handling, and a sense of fun that belied their diminutive dimensions.

The arrival of the Subaru Vivio in 1992 marked another high point in the company's kei car story. Named after the musical term "vivace," meaning "lively," the Vivio exemplified Subaru's knack for combining innovation with practicality. Designed from scratch to take full advantage of revised kei car regulations—now allowing slightly larger dimensions and more advanced safety equipment—the Vivio was offered as a three- or five-door hatchback with an all-aluminum, 658cc four-cylinder engine. The car's lightweight chassis, precise steering, and superb visibility made it an ideal city companion, while the interior was spacious and thoughtfully laid out, featuring folding rear seats and clever storage solutions.

The Vivio stood out for its embrace of advanced engineering in the kei class. The supercharged Vivio RX-R variant, introduced in 1992, was a technological marvel: a 64-horsepower engine, full-time all-wheel drive, four-wheel independent suspension, and available continuously variable transmission (CVT) all in a car less than three and a half meters long. On paper, these features seemed almost extravagant for a kei car, but in practice they showcased Subaru's commitment to giving even its smallest models the same attention to performance and safety found in its flagship products. The Vivio RX-R quickly gained a cult following among young drivers and motorsport fans, competing successfully in one-make racing series and even the infamous Paris–Dakar Rally.

Throughout the kei car era, Subaru remained deeply attuned to the needs of urban Japanese life. Parking was a perennial challenge, and Subaru's kei cars were engineered with impressively tight turning circles, slim pillars for easy maneuvering, and large windows for excellent visibility. Fuel economy was prioritized, with many models delivering over 20 kilometers per liter—critical for families on tight budgets and daily commuters facing ever-rising fuel costs. Serviceability was another hallmark: Subaru designed its kei engines for easy access, quick maintenance, and long service intervals, reducing both downtime and ownership costs.

Marketing of Subaru's kei cars often highlighted their suitability for young families, elderly drivers, and women—demographics that increasingly dominated urban mobility in post-bubble Japan. Advertisements portrayed the R-2, Rex, and Vivio not just as cars, but as partners in everyday life: helping with grocery runs, transporting children to school, and supporting small businesses. Their friendly, approachable styling and customizable color options made them feel less like utilitarian machines and more like trusted companions.

Subaru's commitment to the kei segment was also a matter of national identity and social responsibility. In a country facing mounting environmental and traffic challenges, kei cars played a vital role in reducing congestion, emissions, and dependence on imported oil. Subaru's innovations in lightweight engineering, engine efficiency, and compact packaging contributed to national goals for sustainability and urban liveability. The kei car ethos—simplicity, thrift, and adaptability—mirrored the values of millions of Japanese families navigating rapid social and economic change.

The eventual decline of Subaru's kei car business in the early twenty-first century was less a matter of failure than of strategic realignment. As the brand focused its resources on global platforms and all-wheel-drive vehicles for export markets, it entered partnerships with other Japanese manufacturers (notably Daihatsu) to supply kei cars under badge-engineered arrangements. Even as direct development waned, Subaru's legacy in the kei segment remained secure—testament to decades of innovation, customer focus, and urban problem-solving.

The story of Subaru's kei cars is, at its core, a story of engineering humility and creativity. From the R-2 to the Rex and Vivio, these small cars distilled the brand's defining virtues—reliability, efficiency, clever packaging, and a refusal to compromise on the essentials—into packages perfectly suited for the realities of Japanese urban life. They proved that greatness in automotive design is not always measured in speed or horsepower, but in the ability to meet people's needs with simplicity, style, and ingenuity. Long after the last Vivio rolled off the assembly line, the spirit of Subaru's kei cars endures in the brand's unwavering focus on everyday usefulness, customer trust, and the joy of getting from here to there—no matter how crowded the road might be.

Chapter 17: Green Horizons: Subaru and Environmental Responsibility

As the twenty-first century approached, environmental responsibility ceased to be a niche concern and instead became a defining challenge for automakers around the world. Air quality regulations tightened, climate change concerns came to the forefront, and consumers increasingly sought vehicles that could meet their mobility needs without compromising the planet's future. For Subaru, a company long associated with practicality and resilience, embracing these green imperatives was not merely a strategic adjustment but a natural extension of its engineering philosophy. Through pioneering emission-reduction technologies, cleaner engines, and sweeping sustainability initiatives—including the groundbreaking zero-landfill Indiana plant—Subaru established itself as a leader in environmental stewardship within the automotive industry.

Subaru's journey toward green horizons began with an awareness of its vehicles' environmental impact in the context of Japan's tightly regulated and environmentally sensitive society. The country's mountainous terrain and densely packed cities meant that emissions and noise from cars posed acute challenges, especially in urban centers where the effects were felt most immediately. Japan's Ministry of Transport had begun introducing progressively stricter emissions standards in the early 1970s, well before many Western countries followed suit. Subaru, along with its domestic competitors, was compelled to innovate quickly.

The earliest milestones were achieved through fundamental engineering. In the 1970s, as the world grappled with oil shocks and urban smog, Subaru's engineers worked to optimize the efficiency of their air-cooled and later water-cooled engines. The move to four-stroke engines in the kei car segment, notably in the Rex and later Vivio, was in part a response to stricter exhaust standards. By improving combustion efficiency and introducing cleaner carburetion and exhaust after-treatment systems, Subaru reduced both visible smoke and harmful emissions—a crucial step for continued viability in its home market.

With the development of the Subaru Leone and the Legacy, Subaru adopted new generations of boxer engines that featured improved thermal efficiency and combustion control. The introduction of catalytic converters across the model range further reduced carbon monoxide, hydrocarbon, and nitrogen oxide emissions. Subaru's engineers focused on refining ignition timing, exhaust gas recirculation (EGR) systems, and lean-burn strategies to ensure that vehicles met or exceeded each new emissions threshold, not only in Japan but also in North America and Europe as global export ambitions grew.

A major leap forward came in the 1990s with the widespread adoption of electronic fuel injection (EFI) and on-board diagnostics (OBD). These technologies allowed Subaru to precisely manage fuel delivery and monitor emissions in real time. The result was lower emissions, improved fuel economy, and the ability to swiftly adapt vehicles for different regulatory regimes worldwide. The development of Subaru's "Active Valve Control System" (AVCS), a form of variable valve timing, further optimized combustion for reduced emissions without sacrificing performance. Subaru engines consistently scored well in emissions testing, reinforcing the brand's reputation for balancing engineering excellence with social responsibility.

Subaru's approach to environmental responsibility extended far beyond the engines themselves. In response to growing public and regulatory scrutiny of the automotive industry's broader impact, the company launched an array of sustainability initiatives at its manufacturing facilities. Nowhere was this commitment more evident than at Subaru of Indiana Automotive, Inc. (SIA), the company's U.S. manufacturing plant in Lafayette, Indiana.

Opened in 1989 and expanded repeatedly to meet growing demand, SIA became a global model for green manufacturing. In May 2004, SIA achieved a historic milestone, becoming the first auto assembly plant in the United States—and among the first in the world—to achieve zero-landfill status. Every piece of waste produced during vehicle assembly at SIA was recycled or converted to energy; nothing was sent to a landfill. Achieving this required an extraordinary effort, from redesigning manufacturing processes to changing supplier practices and investing in advanced recycling and waste-to-energy systems.

Employees at SIA were trained to separate waste at the source, and everything from metal offcuts to food scraps from the cafeteria was accounted for. Subaru worked with local partners to ensure that all byproducts—plastics, paper, oil, and even paint sludge—found new life as recycled goods, building materials, or energy sources. The achievement was not just a public relations coup, but a genuine transformation of what automotive manufacturing could mean in the twenty-first century. The Indiana plant's success inspired other Subaru factories and prompted competitors worldwide to raise their own environmental standards.

Subaru's environmental consciousness reached beyond its factories to the communities it served and the landscapes its customers cherished. The company launched multiple initiatives to reduce the environmental impact of its entire product lifecycle, including efforts to design for recyclability, minimize hazardous substances, and use more eco-friendly materials. Subaru joined voluntary programs in Japan and abroad to reduce greenhouse gas emissions and invested in research on alternative powertrains, including hybrid and electric technologies.

The company's advertising and public engagement also reflected its green aspirations. Subaru became a prominent supporter of U.S. National Parks, wilderness preservation, and environmental charities, sponsoring clean-up campaigns and restoration projects. Marketing materials emphasized the compatibility of Subaru vehicles with outdoor adventure and responsible stewardship, positioning Subaru drivers as environmentally conscious explorers rather than mere consumers.

Technologically, Subaru continued to advance engine efficiency and emissions control. The introduction of partial zero emissions vehicle (PZEV) technology in North America during the 2000s set a new benchmark. PZEV Subarus featured advanced catalytic converters, air injection systems, and vapor recovery technologies that dramatically reduced tailpipe emissions, making them among the cleanest gasoline-powered cars available—even meeting stringent California standards. For many years, Subaru offered more PZEV models in the U.S. than any other manufacturer.

As the industry entered the era of electrification, Subaru announced its "Subaru Global Platform," designed to accommodate hybrid and fully electric vehicles alongside traditional powertrains. The launch of the Crosstrek Hybrid in 2018, and the development of the all-electric Subaru Solterra in partnership with Toyota, signaled the company's commitment to meeting both regulatory and customer demand for lower-emissions vehicles while maintaining the brand's core values of all-weather capability and reliability.

Throughout, Subaru has balanced its environmental initiatives with a pragmatic focus on customer needs. Engineers have worked tirelessly to ensure that efficiency gains and emissions reductions do not come at the expense of the performance, safety, or durability that define the brand. The company's cautious yet steady approach to hybrid and electric adoption reflects a desire to deliver real-world solutions, not just headline-grabbing prototypes.

The cumulative impact of Subaru's environmental efforts is visible not only in emissions statistics and sustainability awards but also in the deep loyalty of its customer base. For many Subaru owners—especially in regions like the Pacific Northwest, New England, and Australia—driving a Subaru is an expression of environmental values and a commitment to protecting the landscapes that their vehicles so ably traverse. Subaru's consistent presence in national park parking lots, on mountain trails, and at conservation events underscores the brand's authentic bond with the outdoors and those who cherish it.

In the global narrative of the automotive industry's response to the environmental challenge, Subaru's story stands out for its authenticity, persistence, and results. The company's embrace of green technologies and practices was never just a matter of compliance, but a reflection of a deeper ethos—a belief that cars, and the people who build and drive them, can be part of the solution as well as the problem. From early emissions controls to zero-landfill manufacturing, from clean-burning engines to the promise of electric futures, Subaru's environmental legacy is a testament to what is possible when innovation, responsibility, and care for the planet come together in pursuit of a better tomorrow.

Chapter 18: Facing Adversity: Recalls, Setbacks, and Corporate Reform

No company's journey is without setbacks, and for Subaru—despite its reputation for engineering dependability and customer loyalty—the road to enduring success has been marked by its share of adversity. From technical recalls to internal scandals and sweeping corporate reforms, the brand's evolution has been shaped not only by its innovations and achievements, but also by its capacity to confront problems openly, learn from failures, and emerge stronger. The way Subaru navigated adversity in recent decades stands as an example of resilience and renewal in a relentlessly competitive industry.

The era of modern automotive recalls began in earnest in the late twentieth century, as cars became more complex and consumer safety took center stage. Subaru, like all manufacturers, found itself occasionally grappling with technical issues that required public attention and, sometimes, major corrective actions. Early recalls were relatively modest in scale and scope—issues such as defective seat belt components or faulty emissions controls that could be addressed through dealer fixes. However, as Subaru's production volumes grew and its engineering became more sophisticated, the stakes of quality control and product safety increased.

One of Subaru's most significant technical setbacks occurred in the early 2000s with the emergence of head gasket failures in certain models equipped with the EJ25 series boxer engine. This engine, which powered popular vehicles such as the Legacy, Outback, Forester, and Impreza, gained a reputation for early head gasket leaks that could cause coolant or oil loss. While the fundamental design of the boxer engine contributed to Subaru's celebrated low center of gravity and handling prowess, it also created technical challenges for sealing under high thermal stress. Reports of premature failures and high repair costs threatened to undermine customer trust—especially among long-time owners who prized Subaru for reliability.

Subaru responded by launching extensive technical investigations, extending warranty coverage, and introducing improved multi-layer steel gaskets and revised cooling system protocols in later models. Dealers received updated service bulletins and diagnostic tools, while engineering teams undertook root cause analyses that influenced future engine development. The company's willingness to acknowledge the problem and support affected customers helped to preserve loyalty, even as word spread online and in enthusiast communities.

Another high-profile challenge emerged from within the company itself: the improper inspection and certification of vehicles at Subaru's domestic plants in Japan. In 2017, Japanese regulators revealed that Subaru, along with other automakers, had allowed uncertified workers to perform final vehicle inspections—a violation of strict national protocols intended to ensure quality and safety. The revelation led to a large-scale recall of hundreds of thousands of vehicles, primarily for administrative reasons, and triggered public outcry and media scrutiny.

In confronting this scandal, Subaru's leadership moved quickly to accept responsibility, apologize publicly, and implement a sweeping program of corporate reform. The company voluntarily recalled affected vehicles, strengthened compliance monitoring, and revised training procedures for quality-control staff. Independent audits were conducted to identify systemic weaknesses, while new layers of oversight were installed in manufacturing and inspection lines. Subaru also reorganized its executive structure to emphasize transparency and accountability, underscoring a commitment to ethical practices and continuous improvement.

Financial setbacks have also tested Subaru's resilience. Economic shocks such as the global financial crisis of 2008–2009, currency volatility, and the COVID-19 pandemic placed immense pressure on supply chains, production schedules, and profitability. During these periods, Subaru's relatively compact scale and tightly integrated engineering teams proved to be assets, enabling the company to respond quickly to shifting circumstances. Production was scaled back and costs tightly controlled, but Subaru maintained investment in key research areas, including all-wheel-drive technology, emissions reduction, and electric powertrains. Communication with dealers and customers was prioritized to minimize confusion and maintain trust during periods of product shortage or delayed deliveries.

Subaru has also faced the challenges inherent to rapid growth. As global demand for the brand's vehicles soared in the 2010s—especially in the United States—production capacity was stretched and quality assurance processes were tested. In some years, customer satisfaction surveys indicated that while loyalty remained strong, issues such as delayed deliveries, minor fit-and-finish inconsistencies, and communication breakdowns required urgent attention. Subaru responded by expanding manufacturing at its Indiana plant, adding new supplier partnerships, and investing heavily in staff training and process optimization.

More recently, like all automakers, Subaru has had to navigate the complexities of global recalls related to shared supplier issues, such as the widespread Takata airbag recall that affected millions of vehicles worldwide. Subaru worked in concert with regulators and suppliers to identify affected vehicles, notify owners, and coordinate repairs as quickly as possible, reinforcing the company's longstanding commitment to customer safety.

The cumulative effect of these adversities has been a steady strengthening of Subaru's internal processes and culture. The company invested in advanced quality-control systems, including increased automation, real-time defect tracking, and integrated supplier audits. Cross-functional teams were established to encourage collaboration and early detection of potential problems, while "lessons learned" reviews became standard practice following any recall or production setback.

Subaru's approach to crisis management has also evolved. Where once the instinct may have been to minimize publicity or defer blame, the company increasingly recognizes that openness and proactive engagement are critical to rebuilding trust. Public apologies from top executives, transparent press releases, and direct outreach to affected owners have become hallmarks of Subaru's response strategy. In many cases, these efforts have preserved or even strengthened customer loyalty, with many owners citing Subaru's honesty and support as reasons for repeat purchases.

Perhaps most important has been Subaru's willingness to look inward and reform itself in the wake of adversity. The company's engineers, managers, and frontline workers have come to see setbacks not just as threats, but as opportunities for growth and renewal. The culture of "kaizen"—continuous improvement—has become more deeply embedded, influencing everything from product development cycles to employee training and supplier relations. Subaru's leadership frequently references the lessons of past recalls and crises when making strategic decisions, ensuring that hard-won experience translates into tangible improvements for the future.

In the competitive world of global automaking, even brands with sterling reputations face moments that test their mettle. Subaru's journey through recalls, corporate restructuring, and internal reform has demonstrated a commitment to confronting problems head-on and rebuilding from a foundation of honesty and resilience. These experiences have not only preserved Subaru's bond with its customers but also forged a company better equipped to navigate the complexities of a rapidly changing industry.

The story of Subaru's adversities is thus inseparable from its story of success. It is a reminder that true strength lies not in avoiding difficulties, but in meeting them with clarity, compassion, and a willingness to learn. For Subaru, every challenge faced has become a stepping stone toward greater quality, integrity, and enduring trust—a testament to the brand's capacity for renewal and the deep loyalty it continues to inspire among drivers around the world.

Chapter 19: Globalization and Alliances: The Toyota Partnership

As the twenty-first century unfolded, the automotive world underwent a period of dramatic consolidation and technological upheaval. Fierce global competition, the soaring costs of research and development, and mounting regulatory challenges—particularly in emissions, safety, and electrification—forced automakers to rethink traditional boundaries. For Subaru, a brand proud of its engineering independence and distinct technical identity, adapting to this new landscape required both strategic vision and an openness to alliances. Among these, none would prove more significant than Subaru's deepening partnership with Toyota—an alliance that would help Subaru globalize its operations, share vital technologies, and thrive amid the relentless demands of modern carmaking.

The seeds of the Subaru–Toyota alliance were sown in the early 2000s. At the time, Subaru's parent company, Fuji Heavy Industries (FHI), remained a comparatively small player in the global market, producing around half a million vehicles a year—dwarfed by the output of Japanese giants like Toyota, Honda, and Nissan. Subaru had carved out a loyal customer base in key markets, especially in Japan, North America, and Australia, and had become famous for its signature boxer engines, symmetrical all-wheel drive, and rally-bred models. Yet the realities of globalization—volatile exchange rates, rising R&D costs for safety and emissions, and the need for ever-more sophisticated electronics—made it increasingly difficult for Subaru to compete on its own.

Toyota, meanwhile, was on a trajectory to become the world's largest automaker. Its production scale, financial strength, and expansive technology portfolio made it an appealing partner for many smaller manufacturers. In 2005, Toyota began a strategic investment in Fuji Heavy Industries, initially acquiring an 8.7% stake from General Motors (which was withdrawing from its own alliance with Subaru). Over the years, Toyota would increase its holdings, eventually becoming Subaru's largest shareholder and a de facto guiding partner in several key areas.

At first, the relationship was one of mutual respect and cautious collaboration. Subaru retained operational autonomy and continued to define its own product and brand strategy, while Toyota provided support in specific technical fields and market expansion. Both companies quickly identified opportunities for synergy. Subaru's expertise in all-wheel drive, small-displacement boxer engines, and chassis engineering offered unique value, while Toyota's scale and leadership in hybrid technology, quality management, and supply chain efficiency provided crucial support.

One of the first visible fruits of the alliance was the co-development of the Toyota 86 and Subaru BRZ, a project that began in the late 2000s. The aim was to create an affordable, rear-wheel-drive sports car that would appeal to driving enthusiasts while sharing development costs and production capacity. Engineers from both companies worked side by side, with Subaru providing the boxer engine and much of the platform engineering, while Toyota contributed its direct-injection technology and global marketing reach. The Subaru BRZ and Toyota 86 (also known as Scion FR-S in North America) debuted in 2012, to wide acclaim for their balance, agility, and purity of purpose. The project was a model of successful alliance: each brand retained its identity, but the collaboration delivered a car neither could have produced alone at such a price point.

The success of the BRZ/86 partnership demonstrated the practical benefits of deeper collaboration, paving the way for broader cooperation in areas vital to Subaru's long-term survival. As emissions standards tightened and the market shifted toward electrification, Subaru began to rely on Toyota's advanced hybrid technology. The Crosstrek Hybrid, launched in 2018, was Subaru's first plug-in hybrid and utilized Toyota's well-proven hybrid components paired with Subaru's all-wheel-drive system—an arrangement that maintained Subaru's rugged reputation while delivering competitive efficiency and emissions performance.

Toyota's influence grew as it increased its stake in Subaru to over 20% by 2019. The alliance deepened, with joint committees and shared research projects tackling next-generation powertrains, safety technologies, and connected-car platforms. One of the most significant results was the co-development of the Subaru Solterra and Toyota bZ4X, all-electric crossovers that debuted in the early 2020s. Built on a jointly engineered e-TNGA platform, these vehicles shared batteries, motors, and electronics, while offering distinctive styling, branding, and driving characteristics. Subaru's contribution included its expertise in all-wheel-drive calibration and chassis dynamics, ensuring that even in the era of electrification, Subaru vehicles would retain their characteristic handling and go-anywhere confidence.

Shared platforms and technologies enabled Subaru to keep pace with rivals despite its smaller size and more limited resources. Toyota's purchasing power and global supplier network helped Subaru control costs, manage risk, and maintain consistent quality across increasingly complex models. The partnership allowed Subaru to expand its research into advanced safety systems, including autonomous driving aids and next-generation EyeSight technologies, with access to Toyota's extensive experience in these fields.

Crucially, the alliance preserved the unique aspects of Subaru's brand identity that resonated so deeply with loyal customers. Toyota respected Subaru's independence in product planning, allowing it to continue producing boxer engines, turbocharged performance models, and symmetrical all-wheel-drive systems. Subaru's engineers maintained control over critical ride, handling, and powertrain decisions, ensuring that vehicles like the Outback, Forester, and WRX continued to deliver the driving characteristics that had made the brand famous.

On the global stage, the partnership enabled Subaru to expand its footprint with confidence. North America remained the brand's largest market, but joint ventures and platform sharing allowed Subaru to more effectively serve Europe, China, and emerging regions. Manufacturing cooperation at Subaru's Indiana plant and shared logistics infrastructure improved efficiency and flexibility. The two companies also collaborated on procurement and recycling initiatives, strengthening their commitment to environmental responsibility.

Despite these benefits, Subaru faced the challenges of maintaining its distinctiveness in a world of increasingly homogenized products. Brand loyalists expressed concerns about the dilution of Subaru's core engineering DNA, especially as electrification and platform sharing became more prevalent. Subaru management acknowledged these risks and doubled down on preserving the company's unique technologies and culture, even as it embraced the advantages of the Toyota partnership. Marketing efforts reinforced Subaru's distinct identity, highlighting rally heritage, ruggedness, and customer loyalty, while emphasizing that new technologies would serve these values rather than replace them.

The Toyota alliance also enabled Subaru to navigate periods of industry volatility with greater stability. During global supply chain disruptions, economic downturns, or the COVID-19 pandemic, the partnership's scale and shared resources helped Subaru weather storms that might have threatened its independence in earlier eras. Strategic planning committees, crisis management frameworks, and information-sharing protocols all contributed to a culture of resilience and adaptability.

In the broader context of globalization, Subaru's alliance with Toyota stands as a blueprint for how a small, specialized automaker can survive and thrive in a world dominated by giants. By forging a relationship rooted in mutual respect, shared vision, and complementary strengths, Subaru maintained the freedom to innovate while gaining access to the tools needed for long-term competitiveness. The partnership has produced tangible benefits: new products, cleaner technologies, improved quality, and a more secure future.

Looking ahead, the Toyota alliance positions Subaru to confront the challenges and opportunities of a rapidly changing industry. Whether through electrification, automation, or new models of mobility, Subaru is better equipped to adapt while remaining true to its identity. The brand's journey from independence to partnership is not a story of surrender, but one of evolution—proof that with the right alliances, even the most distinctive voices in the automotive world can find new ways to be heard on the global stage.

Chapter 20: Safety First: Subaru's Leadership in Crash Protection

From its earliest years, Subaru's identity has been shaped by a strong sense of responsibility to its drivers and passengers. This ethic of care became a defining trait as the brand moved from simple, rural workhorses to sophisticated global models. Nowhere is this commitment more visible than in Subaru's leadership in automotive safety—a legacy built on pioneering crash protection, groundbreaking driver-assist technologies, and a relentless pursuit of safer roads for all. Through decades of innovation and discipline, Subaru's dedication to occupant protection and accident avoidance has not only saved lives but also built one of the industry's most loyal customer followings.

Subaru's journey toward safety leadership began with foundational engineering decisions. The adoption of the horizontally opposed boxer engine, for example, was not just about balance and performance; its low, compact design allowed the engine to sit beneath the passenger compartment's crash structure. In the event of a frontal collision, the boxer engine could slide beneath the cabin rather than intruding into it—a layout that proved superior to many upright inline engines in real-world impacts. This was just one early illustration of Subaru's habit of engineering with safety in mind from the outset.

As crash testing and regulatory requirements advanced in the 1970s and 1980s, Subaru's engineers responded by reinforcing passenger cells and integrating deformable crumple zones. The company studied crash data and adapted best practices from both Japan and international markets, even as many rivals focused primarily on cost or style. Features such as energy-absorbing steering columns, seatbelt pre-tensioners, and improved door reinforcement became standard long before government mandates required them.

The 1990s saw a new level of sophistication as computer-aided design enabled more comprehensive crash simulations. With models like the Legacy and Impreza, Subaru systematically improved the structural rigidity of chassis and optimized airbag deployment to cover a wider range of impact types. Side-impact protection, anti-whiplash seats, and the use of high-tensile steel in key areas became standard features, while ongoing crash testing—both in-house and by third parties—allowed Subaru to constantly refine its approach. The company's engineers were often quick to implement lessons from real-world accident investigations, treating customer feedback and crash reports as invaluable tools for improvement.

Subaru's investments in safety began to pay visible dividends as global crash rating programs matured. In the United States, the Insurance Institute for Highway Safety (IIHS) and National Highway Traffic Safety Administration (NHTSA) launched comprehensive testing regimes in the 1990s and 2000s. Subaru models rapidly distinguished themselves, earning top marks and "Top Safety Pick" or "Top Safety Pick+" designations across the lineup. In Japan, Subaru achieved leading scores in the Japan New Car Assessment Program (JNCAP), while in Europe, the brand performed strongly in Euro NCAP testing. These third-party endorsements proved instrumental in convincing safety-conscious buyers—especially families, educators, and medical professionals—to trust Subaru with their transportation needs.

A major turning point arrived with the debut of Subaru's EyeSight Driver Assist Technology in 2008. Developed in-house, EyeSight was among the world's first production camera-based safety suites, leveraging stereoscopic cameras mounted near the rearview mirror to monitor traffic, detect obstacles, and identify lane markings. EyeSight's suite of features included adaptive cruise control, pre-collision braking, lane departure warning, and lead vehicle start alert. Unlike radar-based competitors, EyeSight's vision system could recognize and respond to a wide variety of objects and conditions, including pedestrians and cyclists. Subaru's approach prioritized real-world usability and consistent performance in the challenging weather conditions where many of its vehicles excelled.

The rollout of EyeSight across the global lineup marked Subaru as a leader in active safety and accident avoidance. Independent studies found that EyeSight-equipped Subarus were involved in significantly fewer rear-end collisions and pedestrian accidents compared to similar vehicles without the technology. Over time, the system's sophistication increased, with the addition of lane-keeping assist, reverse automatic braking, blind-spot monitoring, and driver attention alerts. Updates were delivered regularly to improve accuracy and broaden the range of situations in which EyeSight could intervene.

At the same time, Subaru invested heavily in passive safety. The company's Global Platform, introduced in the late 2010s, was engineered from the ground up to maximize crash protection. The platform's ring-shaped reinforcement frame created a continuous safety cell around occupants, while increased use of high-strength steel and multi-path load distribution ensured energy from impacts was absorbed and redirected away from the cabin. Real-world crash data and computer simulations informed the placement of airbags, sensors, and seat structures, allowing Subaru vehicles to perform consistently well even in the most challenging crash scenarios.

Subaru's safety philosophy also extended to the needs of pedestrians and other road users. The brand implemented impact-absorbing front structures, energy-dissipating wipers and hinges, and hoods designed to reduce injury risk in the event of a collision with a pedestrian or cyclist. In markets where child safety is a paramount concern, Subaru vehicles were designed for easy installation of car seats and featured rear door locks, window switches, and alert systems to protect young passengers.

The cumulative effect of Subaru's safety leadership has been profound. In crash tests conducted by IIHS, NHTSA, and JNCAP, Subaru models consistently achieve top-tier results year after year. The brand's reputation for safety is reinforced by real-world outcomes; insurance industry data and independent studies frequently cite Subaru vehicles as among the least likely to be involved in fatal or severe-injury crashes in their classes. This record has translated into strong resale values and deep customer loyalty, with many Subaru owners citing safety as the single most important reason for repeat purchases.

Subaru's commitment to safety is also a key factor in its appeal to niche markets and regions where weather or road conditions make accident risk higher. In snowbelt states in the U.S., mountain regions of Australia, and northern Europe, families, teachers, doctors, and emergency responders have long chosen Subaru for its reputation as the safest way to navigate challenging roads. This relationship is self-reinforcing; Subaru listens closely to these communities, soliciting feedback and integrating lessons learned into future designs.

Even as the industry moves rapidly toward electrification, automation, and new forms of mobility, Subaru's emphasis on safety remains central. The company continues to refine both passive and active protection systems, exploring new technologies such as driver monitoring, cross-traffic alert, and autonomous emergency steering. Partnerships with Toyota and other industry leaders have accelerated the development and rollout of next-generation safety features, while rigorous internal standards ensure that Subaru's signature protection extends to every new platform.

In the end, Subaru's leadership in crash protection is not just a matter of hardware and software, but a culture of care that permeates the entire organization. Engineers, designers, and managers share a common goal: to return drivers and passengers safely to their homes and families, every time. This ethos has become a pillar of Subaru's brand, shaping everything from product development to marketing and dealer training. In a world where safety is increasingly recognized as the ultimate measure of automotive excellence, Subaru's steadfast dedication stands as both a competitive advantage and a moral imperative—proof that, for Subaru, safety is never optional, but always first.

Chapter 21: Subaru's People: Dealers, Engineers, and Loyalists

Behind every Subaru that has ever conquered a snowy pass, survived a city commute, or claimed a rally victory lies a remarkable human story. Subaru's ascent from an experimental automaker in postwar Japan to a global symbol of reliability and adventure is not just a tale of technical achievement or marketing acumen—it is a testament to the collective efforts, values, and vision of the people who shaped, sold, and championed the brand. From the engineers laboring over boxer engines in Gunma to the family-owned dealerships that dot the rural American Northeast, and from club enthusiasts in Australia to generations of fiercely loyal customers, Subaru's culture has always been defined by a unique sense of purpose, humility, and connection. The spirit of Subaru is not just found under the hood; it is embodied in the hands and hearts of those who build, sell, and drive its cars.

The engineers at Subaru, working within Fuji Heavy Industries and later Subaru Corporation, have long occupied a central role in shaping the company's ethos. Many came from humble beginnings, often with backgrounds in aviation, as Subaru's earliest technical teams did. The company's first decades were marked by a spirit of experimentation and frugality, as engineers designed vehicles like the 360 and Sambar van for a nation still recovering from war. Their innovations were rarely about chasing records or prestige; rather, they sought to solve practical problems—how to carry a family safely through crowded cities, how to endure Japan's punishing winters, and how to stretch every liter of fuel.

Subaru's engineering culture grew increasingly distinct as the company pursued unconventional technical paths. The decision to persist with the horizontally opposed (boxer) engine and symmetrical all-wheel-drive systems—technologies considered complex and expensive by many rivals—was rooted in an insistence on stability, safety, and real-world drivability. Teams worked late into the night in Utsunomiya and Gunma, testing prototypes on rain-slicked roads and snowbound tracks, obsessed with incremental improvements in reliability and performance. Their pride was not in high-profile launches but in the quiet satisfaction of solving problems that customers might never even know existed.

The same practical ingenuity extended to Subaru's manufacturing plants. Workers on the lines at Ota, Yajima, and later the zero-landfill Indiana plant in the United States took personal responsibility for quality, often rotating jobs and participating in "kaizen" (continuous improvement) circles to suggest and implement process changes. Many of these workers and engineers stayed with the company for decades, their institutional memory and craftsmanship contributing to Subaru's reputation for robust, long-lasting vehicles.

If Subaru's engineers built the cars, it was the dealer network that brought the brand's values to the public. Unlike the massive dealership conglomerates that typified the American automotive landscape, Subaru's early dealer network was largely comprised of family-run businesses, many located in rural areas and snowbelt states—Vermont, Colorado, Minnesota, and Oregon among them. These dealers were often car enthusiasts themselves, drawn to the brand by its underdog appeal and practical virtues. They knew their customers personally, remembered birthdays and service histories, and sometimes went so far as to deliver a replacement car to a family stranded in a blizzard or drive across state lines to source a rare part for an aging Legacy.

Subaru dealers became trusted fixtures in their communities, building loyalty not with flashy showrooms but through relationships, transparency, and a commitment to customer satisfaction. Stories abound of Subaru dealers going above and beyond during natural disasters—offering shelter during storms, loaning all-wheel-drive cars to rescue teams, or even opening their service bays after hours for stranded motorists. These gestures, while often unpublicized, created bonds of trust and goodwill that translated into generations of repeat customers.

Subaru's approach to sales and service fostered a distinctive customer experience. The company's advertising eschewed celebrity endorsements and aspirational luxury in favor of authenticity and practical value. Slogans like "We built our reputation by building a better car" and "Love. It's what makes a Subaru, a Subaru" resonated with buyers seeking honesty and reliability. The company listened closely to dealer feedback, often adapting product features, marketing, and inventory to local conditions. For example, models with heated seats, larger batteries, and block heaters were prioritized in snowy regions, reflecting an attention to the real-life needs of Subaru drivers.

Perhaps nowhere is Subaru's culture more visible than in the passionate, multigenerational community of owners and enthusiasts who identify with the brand's blend of durability, safety, and subtle performance. Across North America, Australia, New Zealand, Europe, and Japan, Subaru owners have formed clubs, online forums, and social media groups that celebrate every aspect of the brand. Events like Subiefest in the United States, the Subaru Owners Club meets in the United Kingdom, and grassroots rally weekends in Australia draw thousands of fans each year. At these gatherings, WRX and STI drivers swap tuning tips with Outback owners, families display their high-mileage Foresters adorned with national park stickers, and stories are exchanged of Subarus that survived floods, blizzards, and cross-country road trips.

The loyalty of Subaru's customer base is legendary in the industry. It is not unusual to find families who have owned nothing but Subarus for decades, their driveways filled with a succession of wagons, sedans, and crossovers in varying states of age and adventure. Many owners anthropomorphize their cars, giving them nicknames, documenting their journeys online, and even holding informal "retirement" ceremonies when a well-traveled Subaru is finally replaced. This emotional connection is grounded in the trust that the car will start on the coldest morning, carry loved ones safely home, and endure the rigors of daily life with minimal fuss.

Subaru's fan culture is also shaped by a spirit of inclusion and individuality. The brand's appeal crosses generational, geographic, and social lines, uniting outdoor enthusiasts, pet owners, teachers, engineers, and rally fans under a shared ethos of practicality and adventure. The company has embraced this diversity in its marketing and product planning, supporting LGBTQ+ causes, environmental initiatives, and animal rescue organizations that resonate with its broad customer base.

Engineers at Subaru often attend enthusiast events, listening to feedback and answering technical questions. Dealers frequently sponsor local gatherings or participate in charity drives, deepening community ties. The boundary between the company and its customers is porous—ideas and inspiration flow in both directions, resulting in continuous improvements and a sense of shared ownership. It is not uncommon for design tweaks, accessory offerings, or even new model concepts to emerge from conversations with Subaru's most devoted fans.

For many Subaru employees, working for the company is more than just a job—it is a vocation. Pride in the product is matched by pride in the brand's mission to build safe, reliable, and environmentally responsible vehicles. Engineers and line workers sign commemorative vehicles or participate in annual ceremonies marking production milestones. Dealers often pass their businesses from one generation to the next, while customers return to the same showroom year after year, introducing their children and grandchildren to the Subaru family.

In an era when many automakers have grown distant from their roots and their customers, Subaru's people remain at the heart of its success. Their stories—of perseverance, humility, creativity, and loyalty—illustrate why the brand has endured and flourished against the odds. It is a culture built on shared purpose, mutual respect, and the conviction that a car is more than a machine; it is a companion on life's journey. Through the dedication of its engineers, the integrity of its dealers, and the passion of its loyalists, Subaru continues to chart a path defined not only by what it builds, but by the community it inspires and sustains.

Chapter 22: The WRX Phenomenon: Culture, Clubs, and Community

Few cars in modern history have ignited such devotion or inspired such a broad, interconnected following as the Subaru WRX. What began as a rally-bred, turbocharged sedan for enthusiasts soon blossomed into a global phenomenon—a badge that united drivers across continents and cultures through shared passion, performance, and community spirit. The WRX became more than a car; it became a cultural touchstone, a symbol of individuality, and the beating heart of a grassroots movement that transcended roads, racetracks, and even generations.

The transformation of the WRX from high-performance outlier to cultural icon began in the 1990s, when the first Impreza WRX models arrived in Japan and, soon after, in select international markets. Designed as a homologation special for rally racing, the WRX quickly captured the imaginations of young drivers and motorsport fans with its muscular boxer engine, turbocharged thrust, and legendary all-wheel-drive grip. Its accessible price and everyday practicality meant that a broad spectrum of drivers could experience technology and performance once reserved for the rally elite. The WRX was as comfortable on a gravel backroad as it was on a daily commute—a car equally at home attacking hairpin corners or transporting groceries through a snowstorm.

From the outset, the WRX attracted a dedicated base of enthusiasts who recognized its unique blend of capability and fun. This growing community began to organize, meeting first in parking lots and autocross circuits, then in increasingly sophisticated clubs and online forums. In the United Kingdom, Australia, and North America, WRX owners swapped technical advice, compared dyno charts, and organized informal rallies, their camaraderie built on a shared appreciation for the car's performance and underdog appeal. The launch of the WRX STI, with its rally pedigree and unmistakable blue-and-gold livery, only intensified this sense of connection.

The emergence of the internet in the late 1990s and early 2000s accelerated the WRX's ascent to cult status. Online forums like NASIOC (North American Subaru Impreza Owners Club), ScoobyNet in the UK, and RexNet in Australia became vital hubs for technical support, event planning, and storytelling. On these platforms, novices and seasoned tuners alike exchanged everything from repair tips to tales of epic road trips and racing triumphs. Friendships formed in these digital spaces often translated into real-world gatherings—track days, mountain runs, charity drives, and sprawling meets where hundreds of Subarus would converge in a sea of boxer rumble.

One defining trait of the WRX community was its openness and inclusivity. Unlike some performance car scenes, which could feel exclusive or intimidating, WRX clubs welcomed all backgrounds—men and women, families, students, mechanics, and rally fans. The focus was always on shared passion, learning, and mutual respect. For many owners, joining a Subaru club or online forum was as much about finding friends as it was about modifying a car. The community's willingness to help—whether lending tools to a fellow member, sharing a part, or organizing a roadside rescue—became legendary, reinforcing the notion that owning a WRX was entry into a wider family.

Grassroots motorsport played a critical role in the WRX's rise. Autocross events, time attack competitions, and local rally series became proving grounds for both the car and its drivers. The WRX's balance, turbocharged power, and robust all-wheel-drive made it a formidable competitor at all levels of amateur motorsport. Many owners took their first steps into racing behind the wheel of a WRX, confident that their car could handle the abuse and deliver thrills in equal measure. As rallycross and gravel events spread globally, the WRX and STI became standard-bearers, racking up class wins and championships, and bringing a taste of professional rallying to everyday enthusiasts.

The WRX's influence stretched into pop culture and the digital age. Its unmistakable silhouette, flared fenders, and signature hood scoop featured prominently in racing video games, anime, and Hollywood films, further fueling its mystique among younger audiences. Movies like "Baby Driver," the "Fast and Furious" franchise, and games such as "Gran Turismo" and "Forza Motorsport" brought the WRX to living rooms around the world, turning it into an aspirational icon for a generation raised on pixels and pixels per second.

In the real world, WRX meets became destination events. The largest of these, such as Subiefest in California and Boxerfest in the U.S. Mid-Atlantic, drew thousands of cars and fans each year, featuring everything from vendor showcases and autocross to charitable fundraisers and family activities. Internationally, similar gatherings in Japan, the UK, and Australia fostered a spirit of cross-cultural connection—Subaru fans from different continents trading stories, rare parts, and track tips. Manufacturers and dealers recognized the value of these communities, often sponsoring events or sending engineers to interact with owners and gather direct feedback.

The aftermarket industry flourished in the wake of the WRX's popularity. Performance tuners, suspension specialists, and parts suppliers competed to serve the growing demand for customization. Whether chasing rally-style upgrades, show-quality builds, or daily-driver reliability, WRX owners could personalize nearly every aspect of their cars. Some enthusiasts transformed their Subarus into fire-breathing track weapons, while others focused on aesthetics, overland builds, or subtle refinements for daily use. The variety and creativity on display within the community became a hallmark of the WRX phenomenon.

Amid all the horsepower and spectacle, the most enduring legacy of the WRX is the camaraderie and sense of belonging it has fostered. Owners frequently recount stories of being "waved down" by fellow Subie drivers, stopping to help strangers on snowy roads, or forming friendships that lasted far beyond the cars themselves. For many, the WRX is a vessel for experiences—road trips, midnight wrenching sessions, late-night conversations, and shared victories at the track. The blue-and-gold badge became a passport to a world where performance and practicality, individuality and community, could coexist.

Even as the automotive world evolves, and as the WRX itself adapts to new technologies and environmental standards, its community endures. New generations of enthusiasts continue to discover the magic of turbocharged boxer engines, symmetrical all-wheel drive, and the warm welcome of WRX clubs and forums. The culture remains vibrant, adaptive, and fiercely loyal—a testament to the lasting impact of a car that began as a rally tool and became a global movement.

In the story of Subaru, the WRX stands as a powerful symbol of what can happen when engineering excellence meets human connection. It is not just a car, but a catalyst for friendships, adventures, and lifelong memories. Through clubs, meets, forums, and grassroots motorsport, the WRX has given thousands a sense of belonging and identity—proof that sometimes, a car can build a community as enduring as any machine on the road.

Chapter 23: Evolution in the Electric Age: Hybrids, EVs, and the Future

As the automotive industry entered the twenty-first century's third decade, the winds of change began blowing harder than ever before. Concerns over climate change, stricter emissions regulations, shifting consumer attitudes, and technological revolutions all converged to transform how the world thought about transportation. For Subaru—a company known for engineering consistency, loyalty to the boxer engine, and an image of rugged, all-wheel-drive adventure—this transition posed both unprecedented challenges and remarkable opportunities. How could a brand steeped in combustion engineering and rally legend chart a new course in the era of electrification, while preserving its unique identity and ensuring relevance for future generations?

The first seeds of Subaru's electric journey were planted in the early 2000s, long before battery-electric vehicles became a staple of every carmaker's portfolio. Japanese automakers, guided by government policy and early environmental concern, were among the first to experiment with hybrid drivetrains and alternative fuels. Subaru, still a comparatively small manufacturer, initially focused on incremental improvements to combustion efficiency and emissions controls. Engineers refined engine management, introduced variable valve timing, and made widespread use of partial zero-emissions vehicle (PZEV) technologies—especially in North America, where Subaru's commitment to clean air helped build a loyal customer base.

But by the 2010s, it was clear that more dramatic change was needed. Around the world, cities began banning internal combustion engines from their centers, and major markets such as China and the European Union set aggressive targets for zero-emissions sales. At the same time, electric vehicle (EV) technology matured rapidly: batteries grew cheaper and more capable, charging infrastructure expanded, and consumer skepticism began to fade as early adopters shared positive experiences. For Subaru, this was both an existential threat and an open invitation to innovate.

Recognizing that it could not go it alone, Subaru leveraged its deepening partnership with Toyota—the world's leading hybrid manufacturer and a pioneer in scalable electrification. This collaboration gave Subaru access to proven hybrid components, advanced battery management, and the economies of scale that small automakers alone could not achieve. The first visible result was the Subaru Crosstrek Hybrid, launched in North America in 2018. The model combined Subaru's familiar symmetrical all-wheel drive and rugged crossover body with a plug-in hybrid powertrain based on Toyota's Hybrid Synergy Drive technology.

The Crosstrek Hybrid was engineered for versatility. Its lithium-ion battery pack allowed for short electric-only commutes—ideal for city drivers—while the efficient gasoline engine and standard all-wheel drive maintained Subaru's core values of go-anywhere capability and reliability. Though its electric range was modest compared to full EVs, the Crosstrek Hybrid's balance of practicality, economy, and all-weather confidence was enough to appeal to Subaru loyalists and eco-conscious buyers alike. Its launch marked Subaru's public commitment to electrification, and set the stage for more ambitious projects to follow.

The next chapter unfolded with the Subaru Solterra, a fully electric SUV developed in close partnership with Toyota and introduced in the early 2020s. The Solterra—its name a portmanteau of the Latin words for "sun" and "earth"—represented Subaru's boldest move yet into the electric future. Built on the new e-TNGA platform (Toyota's modular EV architecture), the Solterra combined a flat, skateboard-style battery layout with dual electric motors and advanced electronic all-wheel drive. Engineers from Subaru were deeply involved in tuning the system for traction and stability, ensuring that the Solterra delivered the driving confidence expected from any vehicle wearing the six-star badge.

Solterra's launch was more than just a product release; it was a signal that Subaru intended to compete on the front lines of the electric revolution. The vehicle featured state-of-the-art safety systems, over-the-air software updates, a suite of connected services, and a design that blended futuristic cues with Subaru's traditional emphasis on practicality and adventure. Spacious, efficient, and capable of traveling hundreds of kilometers on a single charge, the Solterra expanded the brand's appeal to new markets while providing a zero-emissions alternative for existing Subaru enthusiasts.

At the same time, Subaru announced comprehensive plans for further electrification across its lineup. The company set targets for increasing the share of hybrid and electric vehicles in global sales, pledged to achieve carbon neutrality at its manufacturing facilities by mid-century, and expanded investments in renewable energy and sustainable supply chains. Subaru began exploring the use of recycled materials in interiors, experimenting with new lightweight composites, and investigating solid-state battery technologies that promise even greater range and performance.

But electrification at Subaru has never meant abandoning the traits that made the brand distinctive. The challenge, as articulated by company leadership and engineers, was to ensure that future models—whether hybrid, plug-in, or fully electric—remained true to Subaru's core values: safety, durability, and go-anywhere capability. This focus has shaped every aspect of the electric transition. For example, Subaru's engineers worked to ensure that electric all-wheel-drive systems matched or exceeded the performance of mechanical ones, that ground clearance and cargo space were not sacrificed, and that vehicles were tested for extreme weather reliability in snow and ice—conditions where Subarus are often at their best.

Customer engagement has also played a crucial role in Subaru's electric evolution. The brand has actively sought feedback from its enthusiastic owner base, learning from the early adopters of the Crosstrek Hybrid and Solterra. Marketing and outreach efforts have focused on educating buyers about the realities of EV ownership, addressing concerns about charging infrastructure, range, and long-term durability. Dealers have undergone specialized training to service and support electric drivetrains, ensuring a smooth transition for customers accustomed to traditional Subarus.

Subaru's gradual, measured approach to electrification reflects both the brand's engineering conservatism and its desire to meet real-world customer needs. Rather than rushing to electrify every model overnight, Subaru has prioritized meaningful advances—balancing regulatory demands with reliability, cost, and the expectations of a fiercely loyal customer base. The company has also continued to invest in combustion engines for markets and applications where electrification remains impractical, such as remote rural areas or developing economies.

Looking ahead, Subaru's strategies for the electric age involve ongoing partnerships, investment in new technologies, and a continued focus on its unique market position. By working closely with Toyota and exploring collaborations with other tech leaders, Subaru aims to accelerate the rollout of next-generation EVs, develop proprietary battery and all-wheel-drive technologies, and expand its role in the rapidly shifting mobility ecosystem. Research into vehicle-to-grid (V2G) integration, renewable energy storage, and smart mobility solutions reflects the brand's intention not just to survive, but to thrive in a world where the definition of "automaker" is constantly evolving.

The story of Subaru's entry into electrification is still being written. As battery technology improves, charging networks expand, and customer preferences shift, the brand will face new choices about how to blend tradition with innovation. But if history is any guide, Subaru will continue to do what it has always done best: focus on the real-world needs of its customers, deliver honest engineering, and build vehicles that inspire loyalty, trust, and adventure—no matter what powers them. In the electric age, Subaru's challenge and opportunity is to prove that its spirit—the resilience, ingenuity, and sense of connection that made it a global icon—will endure, even as the world and its roads change forever.

Chapter 24: Enduring Identity: What Makes a Subaru a Subaru

Across more than seven decades, amid shifting technologies, changing consumer demands, and the ceaseless evolution of the global auto industry, Subaru has managed to preserve something rare—a distinctive, instantly recognizable identity. While many brands have bent to the prevailing winds of fashion or merged into the mainstream, Subaru has repeatedly chosen a different path, forging its own reputation on the strength of unique engineering, authentic marketing, and deeply held values. The answer to the question, "What makes a Subaru a Subaru?" lies not in any one feature or advertising slogan, but in the sum of choices—technical, cultural, and emotional—that together form a resilient, lasting brand appeal.

The roots of Subaru's identity stretch back to its earliest years, when Fuji Heavy Industries shifted from building aircraft to manufacturing cars. The lessons of those aviation origins—precision, safety, efficiency, and a respect for engineering rigor—shaped every vehicle that followed. Unlike many Japanese contemporaries, Subaru set out not to imitate but to innovate. Early cars such as the 360, Sambar, and R-2 were small, light, and practical, engineered to meet the needs of a rapidly urbanizing society. Yet they also bore the mark of technical ambition: a willingness to try rear-mounted engines, to refine front-wheel-drive and four-wheel-drive layouts, and to build vehicles not just for the city but for Japan's often unforgiving countryside.

As Subaru matured, two key engineering choices cemented its place in automotive history. The first was the horizontally opposed (boxer) engine, adopted in the 1960s and made standard across the lineup in the following decades. The boxer's low center of gravity and compact dimensions offered unique handling and safety benefits—virtues that would become central to Subaru's pitch to drivers. The second was the brand's early and persistent focus on all-wheel drive. Initially an option on utilitarian wagons and vans, Subaru's symmetrical all-wheel-drive system evolved into a core technology, setting the brand apart as the world's leading manufacturer of affordable, everyday AWD cars and crossovers.

This technical foundation allowed Subaru to develop vehicles that felt different from the competition—stable, balanced, confidence-inspiring in all weather and on all surfaces. The company's willingness to sacrifice a degree of mass-market simplicity for these benefits led to quirks—longitudinal engines, asymmetrical exhausts, unconventional packaging—but also gave Subarus a mechanical honesty and character that inspired loyalty in owners. When rivals marketed style or speed, Subaru offered security and trust; when others pursued lowest-common-denominator costs, Subaru doubled down on ruggedness and reliability.

Yet identity is shaped by more than engineering. Subaru's marketing and brand communication have always projected a tone of authenticity, humility, and community. In Japan, the brand cultivated a following among rural drivers, craftsmen, and outdoors enthusiasts—people who valued durability over status. In North America, Subaru staked its reputation in the snowy, mountainous, and rural corners of the continent. Advertising campaigns avoided celebrity endorsements and instead featured real families, teachers, adventurers, and pet owners, presenting Subarus as partners in life's journeys rather than objects of aspiration.

Subaru's long-running association with environmental stewardship, national park partnerships, and social causes reinforced the idea that owning a Subaru was about more than transportation—it was a choice aligned with values. By supporting LGBTQ+ initiatives, animal rescue groups, and disaster relief efforts, Subaru built a brand persona of inclusiveness, kindness, and practical action, further deepening emotional bonds with customers.

Central to Subaru's enduring identity is its relationship with its owners and fans. Unlike many automakers, Subaru never distanced itself from grassroots enthusiasm or sought to control how customers experienced the brand. WRX and STI clubs, Outback and Forester overlanding groups, and family-focused events created a participatory, welcoming community. Subaru listened and responded, integrating feedback into model updates, accessories, and marketing strategies. The company's network of dealers—many family-owned and rooted in local communities—extended this relationship, building trust through transparency and personal service.

Subaru's products have also remained consistent in their appeal to core values: safety, utility, and real-world performance. The company invested early and deeply in crash protection and driver-assist technologies, long before they became industry norms. Models like the Outback, Forester, and Crosstrek were designed for practicality—ample cargo space, easy entry and exit, high ground clearance—rather than fleeting fashion. Even performance icons such as the WRX and BRZ balanced everyday usability with excitement, ensuring that customers never had to choose between thrill and function.

Another defining characteristic has been Subaru's willingness to embrace the unconventional and champion the underdog spirit. While not the largest, fastest, or flashiest, Subaru built a following by staying true to its principles and refusing to chase every passing trend. This approach cultivated deep loyalty and word-of-mouth advocacy. Subaru became the car for those who cared more about what a vehicle could do than how it looked in the valet line—a badge for teachers, nurses, environmentalists, and rally fans alike.

Importantly, Subaru's identity has proven adaptable yet resilient in the face of change. The company weathered economic downturns, technical setbacks, and the shift to electrification by grounding each decision in its core philosophy. The adoption of hybrid and electric technology, for example, was undertaken not to chase headlines but to deliver genuine, reliable progress. Even in the era of the Solterra and future EVs, the brand's priorities—traction, safety, versatility—remain firmly intact.

The enduring appeal of Subaru is thus a product of coherence—every aspect of the brand, from product planning to advertising, from dealer culture to engineering decisions, reinforces a singular vision. It is a vision that values honesty over hype, substance over style, and community over conquest. What makes a Subaru a Subaru is not just the way it handles a mountain road or shrugs off a snowstorm, but the sense of trust, belonging, and usefulness it offers its drivers.

In a crowded, ever-changing automotive world, this steadfastness has made Subaru not only distinctive but enduringly attractive. To own a Subaru is to join a community with shared values and a shared journey—a brand that, by remaining true to itself, has managed to remain relevant and beloved across generations. As the industry races toward the unknowns of the future, Subaru's enduring identity offers a model for how conviction, clarity, and care can outlast any trend, ensuring that the spirit of the six stars continues to shine for years to come.

Chapter 25: From Snowdrifts to Street Legends: The Subaru Legacy

As Subaru's journey unfolds across decades, continents, and cultures, the brand's trajectory reads as one of the most remarkable stories in automotive history. From its first tentative steps as Fuji Heavy Industries, rooted in Japan's postwar recovery, to its place as a global standard-bearer for safety, durability, and performance, Subaru's rise has defied industry convention at every turn. The company's transformation—from a producer of humble kei cars and practical wagons to a builder of rally icons and beloved family vehicles—has left an imprint not only on roads but also on hearts, landscapes, and imaginations around the world.

Subaru's global impact begins in its home country, where postwar engineers, many with backgrounds in aviation, turned their technical discipline and resourcefulness toward making mobility affordable and reliable for a nation in need. Their solutions were ingenious and often unconventional: the Subaru 360, launched in 1958, captured the spirit of the era with its compact dimensions and efficient engine, providing Japanese families with freedom and opportunity. As Japan modernized and urbanized, Subaru adapted, expanding its range with practical, space-efficient vehicles that became trusted companions for millions of daily journeys.

The brand's unique engineering philosophy—anchored by the boxer engine and symmetrical all-wheel drive—emerged in the 1960s and 1970s as a direct response to the challenges of Japan's rural terrain and unforgiving winters. This approach, emphasizing stability, safety, and adaptability, distinguished Subaru from a crowded field of domestic rivals and paved the way for a future where the company's core technologies would become signatures recognized the world over.

Yet it was beyond Japan's borders that Subaru's identity would crystallize and spread. In the snowbound states of the American Northeast and mountainous stretches of the Pacific Northwest, Subaru vehicles earned a reputation for relentless dependability in conditions that humbled less capable rivals. Stories abounded of Subarus that started on the coldest mornings, powered through blizzards, and ferried doctors, teachers, and rescue workers through roads others dared not travel. The company's steady focus on all-wheel drive, long before the technology became mainstream, set Subaru apart as the automaker of choice for those who valued function over flash and reliability over risk.

The rise of the Outback in the 1990s marked Subaru's leap from niche favorite to cultural phenomenon in North America and Australia. By blending the utility of a wagon with the toughness of an SUV, Subaru pioneered the crossover segment and redefined what adventure and practicality could mean for ordinary families. The Outback's image—rugged, ready, and at home on both tarmac and trail—became synonymous with Subaru itself, helping to shape the company's marketing, product planning, and growing international presence.

Meanwhile, in the world of motorsport, Subaru carved out a parallel legacy on rally stages across the globe. The arrival of the Impreza WRX and WRX STI—machines born from the heat of World Rally Championship competition—captured the imaginations of enthusiasts and inspired a vibrant culture of clubs, grassroots motorsport, and digital communities. The thunderous sound of a turbocharged boxer engine and the flash of blue and gold livery became emblems of speed, resilience, and the joy of driving. Victories by legendary drivers such as Colin McRae, Richard Burns, and Petter Solberg not only filled Subaru's trophy cabinets but also cemented the brand's place in the annals of performance history.

These twin pillars—everyday dependability and sporting glory—fed into a singular global identity. Subaru vehicles became a fixture not only in snowy driveways and rally paddocks, but also in the consciousness of generations of drivers who valued authenticity, humility, and adventure. The company's enduring commitment to safety and environmental responsibility broadened this appeal, building trust through crash-test excellence, zero-landfill manufacturing, and a genuine engagement with causes that mattered to its customers.

Subaru's ability to remain distinct amid waves of consolidation and change owes much to its people—the engineers who championed unorthodox solutions, the dealers who became community fixtures, and the loyalists who proudly flew the six-starred flag. The relationship between Subaru and its owners has always been more partnership than transaction, built on shared values and a mutual appreciation for substance over spectacle. Subaru drivers have formed an extended, inclusive family: from WRX club meets and Outback adventures to family driveway traditions and viral stories of high-mileage Legacies that refuse to quit.

In the age of electrification, connectivity, and mobility disruption, Subaru's legacy is defined not by clinging to the past, but by evolving while holding true to the principles that made it unique. The transition to hybrids and electric vehicles, driven by partnerships such as the one with Toyota, reflects a continued focus on real-world performance, reliability, and the customer's needs above all else. Even as new models emerge and new challenges arise, Subaru's vision remains grounded in honest engineering, environmental stewardship, and a sense of social responsibility.

Ultimately, Subaru's story is one of transformation—of a company that moved from the margins to the mainstream without ever losing its soul. From the snowdrifts of rural Japan and North America to the street legends of rally lore and city commutes, Subaru has remained a symbol of resilience, ingenuity, and belonging. Its vehicles, whether crossing mountain passes or lining suburban school lots, tell a story of lives lived with confidence, curiosity, and connection.

What endures is not just a portfolio of products or a list of technical achievements, but a legend built on trust, loyalty, and the courage to go a different way. In an industry often defined by change for its own sake, Subaru's legacy is proof that enduring success is possible when a company listens, adapts, and, above all, remains true to the qualities that make it unique. The six stars shine on—reminders of a brand that, from humble beginnings, became nothing less than a global street legend, beloved wherever roads are long, weather is wild, and drivers seek something more than just a way from A to B.

Epilogue

Subaru's journey is, in many ways, just beginning. The story of a brand that sprang from postwar adversity and aerospace ingenuity, grew into a quiet champion of practicality and resilience, and ultimately earned global recognition for both everyday reliability and rally-bred excitement, continues to unfold with each passing year. As the world turns inexorably toward new technologies and uncharted terrain, Subaru finds itself standing at another crossroads—one that demands not only innovation, but the wisdom to carry forward the best of what has come before.

In today's automotive landscape, where electrification, automation, and digital connectivity drive relentless change, Subaru's enduring values—trust, safety, authenticity, and a sense of community—offer rare continuity. These are qualities that cannot be engineered overnight or summoned by marketing alone; they are the product of decades spent listening to customers, solving real-world problems, and refusing to compromise on the essentials. In a market often dominated by short-term trends and shifting allegiances, Subaru's commitment to honest engineering and meaningful purpose has allowed it to retain not just relevance, but a kind of quiet reverence among its owners.

The challenges that lie ahead are as formidable as any Subaru has faced. The shift to electric and hybrid vehicles, the pressure to reduce carbon footprints, and the demands of an interconnected, always-on world require agility and investment on a global scale. The company's alliance with Toyota, its investments in electrification, and its expansion into new forms of mobility all reflect a willingness to adapt without surrendering the character that has always set Subaru apart. The vehicles rolling out of Gunma and Indiana today carry with them not just new technologies, but the collective experience and ethos of generations—engineers, factory workers, dealers, and drivers alike—who have shaped the brand's distinctive path.

Yet Subaru's legacy is not measured only in innovations or sales figures, but in the lives it touches and the journeys it enables. The battered Forester that still soldiers through mountain snows. The WRX whose turbocharged heart beats in time with the dreams of its driver. The Outback loaded for a cross-country adventure, or the Vivio remembered fondly for carrying a family's hopes through city streets. These stories, multiplied by millions, are the living testament to Subaru's enduring connection with its customers—a connection forged not just in showrooms and service bays, but on highways, rally stages, and winding back roads across the globe.

Looking ahead, the task for Subaru is to carry forward this spirit as it embraces new opportunities and overcomes inevitable obstacles. The pressures of regulation, the need for environmental stewardship, and the shifting tastes of a new generation of drivers will shape the contours of the road to come. Subaru's challenge will be to innovate without losing the essence that has made it beloved: engineering that prioritizes safety and confidence, marketing that speaks with honesty, and a community that values authenticity above all.

In a world hungry for substance amid spectacle, Subaru remains a reminder that true progress is not about chasing the latest fad, but about building something that lasts—something that endures through snowdrifts, city traffic, and the test of time. As the company navigates the uncharted paths of the future, the guiding lights remain clear: the six stars that have come to symbolize resilience, ingenuity, and a promise kept.

The story of Subaru is still being written, with each new model and every new generation of driver adding their chapter. The journey continues, propelled by the same conviction and quiet courage that transformed a small Japanese automaker into a name whispered with pride in every corner of the world. For Subaru, the road ahead is as full of promise as it is of memory—a path defined not only by where it leads, but by the spirit with which it is traveled.

Appendix A: Complete Model Chronology

1950s

- **1954 – Subaru 1500 (P-1)**: First Subaru car, a prototype mid-size sedan powered by a water-cooled inline-4
- **1958 – Subaru 360**: Iconic kei-class microcar with rear-mounted two-stroke engine, priced for mass mobility

1960s

- **1961 – Subaru Sambar**: Cab-over kei truck/van, beginning long-running commercial vehicle line
- **1966 – Subaru 1000**: First front-engine, front-wheel-drive family car, introducing Subaru's boxer engine layout
- **1969 – Subaru FF-1 Star & FF-1 G**: Compact front-engine models marking Subaru's early steps into various body styles
- **1969 – Subaru R-2**: Successor to 360 with updated two-stroke engine, later upgraded with water cooling and facelift in 1971

1970s

- **1971 – Subaru Leone**: Compact engineered as sedan, coupe, and wagon, first Subaru passenger car with optional four-wheel drive
- **1972 – Subaru Rex**: Kei-class offering evolved from R-2, introducing four-stroke EK engines
- **1977 – Subaru BRAT (Brumby/MV)**: Quirky pickup with rear-facing jump seats to avoid U.S. tariff—turbo version added later

1980s

- **1983 – Subaru Domingo (Libera/Sumo/Estratto)**: Microvan for both family and commercial use
- **1984 – Subaru Justy**: Compact hatchback, also sold internationally with Daihatsu badge
- **1985 – Subaru XT (Alcyone/Vortex)**: Sporty wedge coupe incorporating boxer engine and AWD
- **1988–1992 – Subaru Bighorn**: Large SUV rebadged from Isuzu Trooper
- **1989 – Subaru Legacy (Liberty)**: Launch of mid-size successor to Leone; introduced new EJ boxer engine and AWD

1990s

- **1991 – Subaru SVX (Alcyone)**: Premium grand-touring coupe with 3.3 L flat-6 engine
- **1992 – Subaru Vivio**: Advanced kei-class hatchback with supercharged AWD variant
- **1992 – Subaru Impreza**: Compact designed for performance; later spawned WRX and STI variants
- **1994 – Subaru Impreza WRX/STI**: Turbocharged performance variant that became world-renowned
- **1995 – Subaru Outback**: Raised Legacy wagon that defined the crossover segment
- **1997 – Subaru Forester**: Compact crossover blending Impreza chassis with raised stance

2000s

- **2001–2004 – Subaru Traviq**: Seven-seat MPV based on Opel Zafira
- **2002–2006 – Subaru Baja**: U.S.-only pickup-truck variant of Impreza/Outback
- **2003 – Subaru R-2 (new kei)**: Spiritual successor to original R-2 launched in Japan
- **2005–2014 – Subaru Tribeca**: Three-row crossover US-developed SUV
- **2006 – Subaru Stella**: Kei car similar to Daihatsu Move
- **2008–2018 – Subaru Exiga (Liberty Exiga)**: Seven-seat MPV based on Legacy

2010s

- **2010–2015 – Subaru Lucra**: Kei minivan rebadged from Daihatsu Tanto Exe
- **2010–2016 – Subaru Trezia**: Subcompact MPV rebadged Toyota Ractis
- **2011–Present – Subaru Pleo Plus**: Kei-class hatchback based on Daihatsu Mira
- **2012 – Subaru BRZ / Toyota 86**: Lightweight, rear-wheel-drive sports coupe
- **2012–Present – Subaru Crosstrek (XV)**: Subcompact crossover derived from Impreza
- **2013 – Subaru Chiffon**: Tall kei wagon rebadged Daihatsu Tanto
- **2014–Present – Subaru Levorg**: Sport-wagon derived from Impreza/WRX

- **2018 – Subaru Ascent (Evoltis)**: Three-row midsize crossover for North America

2020s

- **2022 – Subaru Solterra**: First mass-production BEV from Subaru, co-developed with Toyota
- **2022 – Subaru Rex Crossover**: Reintroduced Rex nameplate as kei crossover (rebadged Daihatsu Rocky)
- **2023 – Subaru Levorg Layback**: New variant of Levorg in Japanese market
- **2025 – Subaru Trailseeker**: Electric mid-size crossover (bZ4X touring derivative)

Summary Table (Post-1989 flagship models)

Model	Launch	Technical Milestone
Legacy	1989	EJ boxer engine, AWD, world debut
Impreza	1992	Compact, rally-ready AWD platform
Forester	1997	Pioneered crossover concept
Outback	1995	Launched segment-defining crossover
BRZ	2012	RWD sports car, Toyota joint effort
Solterra	2022	First EV; Toyota e-TNGA platform

This chronological guide surveys Subaru's evolution across kei cars, compact sedans, wagons, rugged crossovers, performance icons, and electric vehicles—marking key launches and engineering milestones.

Appendix B: Motorsport Achievements

A concise record of Subaru's most significant rally and racing milestones that shaped its performance legacy.

🏁 World Rally Championship (WRC) Highlights

- **1990–1992**: Debut and development with Legacy RS – Subaru enters WRC, scoring first podium in 1990; MSG team forms in 1988 with Prodrive collaboration
- **1993**: Impreza 555 Group A debuts; first WRC win at Rally New Zealand with Colin McRae
- **1995–1997**: Subaru wins **three consecutive WRC Manufacturers' Championships (1995, 1996, 1997)** with the Impreza WRC
- **1995, 2001, 2003**: Subaru drivers claim **three WRC Drivers' Championships**—Colin McRae (1995), Richard Burns (2001), and Petter Solberg (2003)
- **46 WRC victories** earned by the Impreza WRC, including Piero Liatti's historic first world rally win for the WRC-spec Impreza (Monte Carlo, 1997)
- **Notable drivers**: Colin McRae, Richard Burns, Petter Solberg, Piero Liatti, Ari Vatanen, Juha Kankkunen, Tommi Mäkinen, Carlos Sainz

---

Rally Milestones & Events

- **1997 Monte Carlo Rally**: Liatti secures the first WRC win for a World Rally Car with the Impreza
- **2000 Season**: Richard Burns wins Rally Portugal and Rally Argentina driving the Impreza S6 WRC '00

- **2004 Rallye Deutschland**: Petter Solberg pilots the Impreza S10 WRC '04 in a high-profile factory-run event

---

⚙️ **Subaru Tecnica International (STI) Achievements**

- **1988**: STI is founded to spearhead WRC efforts
- **1989**: Legacy RS endurance demonstration—three cars average over 145 mph for extended duration
- **Special Editions**: STI introduces limited-run performance models like the 22B, S201–S204 to commemorate rally successes

---

US **Subaru Rally Team USA (ARA & Rallycross)**

- **2001–present**: Subaru Rally Team USA claims **18 of 24 national rally titles**, including **nine straight** from 2011 to 2019
- **Rallycross success**: Team secures national Rallycross championships in **2019** and **2021**
- **2023**: New WRX rally car wins debut in Ojibwe Forests Rally, helping Brandon Semenuk clinch the ARA Championship

---

Other Motorsport

- **Nürburgring 24 Hours**: Competition entries starting in 2005; significant class wins in SP3T (2011)

- **Super GT & Other Series**: Non-WRC racing saw participation in Japan's Super GT (Legacy B4 GT300) and other international endurance events

---

🏆 Summary of Key Racing Accolades

- **Manufacturers'**: 3× WRC titles (1995–1997)
- **Drivers'**: 3× WRC titles (1995 McRae, 2001 Burns, 2003 Solberg)
- **WRC Victories**: 46 wins, first WRC car victory (1997), multiple rally wins by top drivers
- **National Rally Excellence**: Dominance in US rally and rallycross (2001–present)
- **Endurance & GT racing**: Nürburgring class victories and Japanese GT class participation

---

This Motorsport Appendix captures Subaru's performance pedigree—an enduring testament to its engineering prowess, competitive spirit, and role in building a legacy that transcended the rally stage to define the brand itself.

Appendix C: Technical Specifications of Landmark Subaru Models

A concise, data-driven breakdown of Subaru's most iconic models—highlighting engines, drivetrains, and key engineering breakthroughs.

◆ 2005–2021 Subaru WRX STI (VA/old Impreza-based)

- **Engine**: 2.5 L turbocharged EJ-series flat-four
    - US output: 305 hp @ 6000 rpm (MY2008–'18), 310 hp (MY2019–'21)
    - Torque: ~290 lb-ft (393 N·m) at 4000 rpm; S209 special: 341 hp & 330 lb-ft
- **Transmission**: 6-speed TY85 manual, standard center-differential DCCD
- **Drivetrain**: Symmetrical full-time AWD with front/rear LSDs
- **Performance highlights**: 0–60 mph ~5.3 s; Brembo brakes & Sachs dampers

---

◆ 2025 Subaru WRX (VB platform)

- **Engine**: 2.4 L direct-injection turbo boxer four (FA24F)
    - Output: 271 hp; Torque: 258 lb-ft @2000–4800 rpm
- **Transmissions**: 6-speed manual or automatic "Subaru Performance Transmission" (SPT/CVT with paddle shifts)

- **Chassis**: Subaru Global Platform; rear anti-roll bar and adjustable shocks on tS/GT trims

---

◆ **2025 Subaru Outback (6th/7th generation)**

- **Standard engine**: 2.5 L NA boxer four — 182 hp, 176 lb-ft
- **Optional engine**: 2.4 L turbo boxer four — 260 hp, 277 lb-ft; Tow capacity up to 3,500 lb
- **Transmission**: Lineartronic CVT with simulated "speeds" and paddle control; standard AWD
- **Efficiency**: EPA ~26/32 MPG city/highway; 18.5 gal tank ≈ 481–592 mi range

---

◆ **1992–2024 Legacy / Outback Heritage**

- **Engine lineup**: EJ-series boxer engines; turbo 2.0–2.5 L EJ20/EJ25 in GT / XT variants; PZEV variants offered in select markets
- **Drivetrain**: Symmetrical AWD standard across all trims

◆ **1997–present Forester**

- Based on Impreza but with SUV-style chassis, increased ground clearance, and optional turbo engines (EJ25 XT ~210 hp in early 2000s)

◆ **2012 Subaru BRZ / Toyota 86**

- **Engine**: 2.0 L FA20F boxer with direct injection & twin-scroll turbo? *(actually naturally aspirated 2.0 L, 200+ hp range)*
- Rear-wheel-drive, engineered in cooperation with Toyota for balanced chassis and driver engagement

---

◆ **2022 Subaru Solterra (EV)**

- Built on Toyota e-TNGA skateboard platform with dual electric motors and AWD—marking Subaru's first mass-production EV

◆ **Key Engineering Innovations**

Innovation	**Significance**
Boxer engine layout	Low center of gravity & safety benefits in frontal crashes
Symmetrical AWD	Balanced weight distribution, traction in diverse conditions
EJ-series turbos (EJ20/EJ25)	Iconic performance engines with rally heritage
Global Platform	High-strength steel chassis designed for safety & EV compatibility
EyeSight & driver assists	Camera-based safety suite enhancing collision avoidance

---

These specifications reflect Subaru's blend of distinct engineering—from rally-proven powerplants to adaptive EV platforms—demonstrating why these models have become benchmarks in performance, utility, and innovation.

Appendix D: Glossary of Subaru and Automotive Terms

This glossary provides clear definitions and explanations for important Subaru-specific and general automotive terms, acronyms, and engineering concepts referenced throughout the book. It is intended to help both newcomers and seasoned enthusiasts better understand the language and technology of Subaru and the wider automotive world.

All-Wheel Drive (AWD):
A drivetrain system that powers all four wheels of a vehicle simultaneously, improving traction and stability on slippery or uneven surfaces. Subaru's signature Symmetrical AWD system is renowned for its balanced layout and consistent performance.

Boxer Engine:
An internal combustion engine with horizontally opposed cylinders ("flat" configuration). The opposing movement of the pistons provides a low center of gravity and smooth operation. Subaru has used boxer engines in nearly all of its vehicles since the 1960s.

BRZ:
A lightweight, rear-wheel-drive sports coupe developed jointly by Subaru and Toyota (Toyota 86/GR86). BRZ stands for "Boxer, Rear-wheel drive, Zenith."

Catalytic Converter:
An exhaust system component that uses chemical reactions to convert harmful gases from combustion into less polluting substances before they exit the tailpipe.

Crossover (CUV):
A vehicle that combines features of passenger cars and sport utility vehicles (SUVs), typically built on a car platform with higher ground clearance and available AWD. The Subaru Outback is a pioneering example.

CVT (Continuously Variable Transmission):
A type of automatic transmission that provides seamless, stepless gear ratio changes, allowing for smoother acceleration and improved fuel efficiency. Subaru's Lineartronic CVT is widely used across its lineup.

DCCD (Driver Controlled Center Differential):
A technology in performance Subaru models (notably WRX STI) allowing the driver to adjust the front-rear torque split for better handling and traction in different conditions.

Differential:
A mechanical device that splits engine torque and allows wheels to rotate at different speeds, which is essential for smooth cornering. Subaru vehicles may use several types (open, limited-slip, center differential) depending on the model and trim.

EJ/FA/FB Engines:
Series codes for Subaru's families of boxer engines. EJ (introduced 1989) includes many legendary turbocharged engines; FB and FA are newer, with improved efficiency and emissions.

EyeSight:
Subaru's proprietary driver-assist suite, based on stereoscopic cameras mounted near the rearview mirror. Features include adaptive cruise control, pre-collision braking, lane-keeping assist, and more.

Flat-Four / Flat-Six:

Colloquial terms for four- and six-cylinder boxer engines, where "flat" refers to the horizontal orientation of the cylinders.

Forester:

A compact crossover introduced in 1997, blending car-like handling with SUV practicality. Known for its reliability and versatility.

Hybrid Vehicle:

A vehicle with both a conventional gasoline engine and an electric motor, designed to improve fuel efficiency and reduce emissions. Subaru's Crosstrek Hybrid is an example.

Impreza:

A compact car introduced in 1992, available in sedan and hatchback/wagon forms. Basis for performance models such as the WRX and WRX STI.

Kei Car:

A Japanese vehicle category for microcars with limited engine size and exterior dimensions, intended for urban use. Subaru produced several kei models, including the 360, Rex, and Vivio.

Legacy:

Subaru's midsize sedan/wagon line, introduced in 1989, known for its robust engineering and as the platform for the Outback and various rally homologation specials.

LSD (Limited-Slip Differential):

A type of differential that improves traction by limiting the speed difference between wheels. Used on performance Subarus for better grip in demanding conditions.

Lineartronic:

Subaru's proprietary continuously variable transmission (CVT) technology.

Outback:

A raised, rugged wagon introduced in 1995 that helped define the modern crossover segment. Known for its practicality, ground clearance, and "go-anywhere" image.

PZEV (Partial Zero Emissions Vehicle):

A designation for vehicles meeting strict emissions standards, with near-zero evaporative emissions. Subaru was a leader in offering PZEV gasoline vehicles in the U.S.

Prodrive:

A British motorsport and engineering company that partnered with Subaru to develop and run its World Rally Championship (WRC) program from the late 1980s through 2008.

Rally (Stage Rally):

A motorsport discipline in which modified or production-based cars compete on closed public or private roads, often over dirt, snow, or gravel. Subaru's rally successes, particularly with the Impreza WRX, shaped much of its performance reputation.

Rear-Wheel Drive (RWD):

A drivetrain layout where only the rear wheels are powered. Subaru uses RWD exclusively in the BRZ and related Toyota sports coupes.

SI-Drive (Subaru Intelligent Drive):

A driver-controlled system that lets the user select different throttle and transmission response profiles, optimizing for economy, comfort, or performance.

Solterra:
Subaru's first mass-market battery electric vehicle (BEV), developed in partnership with Toyota and based on the e-TNGA platform.

STI (Subaru Tecnica International):
Subaru's motorsport and high-performance division, responsible for developing WRX STI models and overseeing factory racing efforts.

Subaru Global Platform (SGP):
A modular vehicle architecture introduced in 2016, underpinning new models with improved safety, rigidity, and adaptability for electrified drivetrains.

Symmetrical AWD:
Subaru's signature full-time all-wheel-drive system, characterized by a balanced (symmetrical) layout that contributes to predictable handling and stability.

Turbocharger:
A forced induction device that increases engine power by using exhaust gases to spin a turbine, forcing more air into the engine. Subaru popularized turbocharging with the Legacy, Impreza WRX, and Forester XT.

VDC (Vehicle Dynamics Control):
Subaru's electronic stability control system, which helps maintain vehicle control during extreme maneuvers or on slippery surfaces.

WRX (World Rally eXperimental):
Performance-oriented model based on the Impreza, originally developed as a homologation special for rally competition and later produced as a standalone performance car.

About the Author

Jack Mercer is an independent automotive historian and author whose work explores the vehicles, innovators, and industries that defined motoring across generations. His books offer meticulously researched, narrative-driven accounts of iconic models and marques—from mainstream legends to specialist manufacturers—combining technical accuracy with engaging storytelling.

Mercer's writing appeals to both casual enthusiasts and serious collectors, offering a clear, detailed view of the automotive world's evolution. His focus spans performance machines, motorsport-derived engineering, and the cultural impact of cars across global markets.

When not writing, Jack Mercer devotes his time to studying factory archives, attending historic vehicle events, and documenting the stories of the people and machines that helped shape the road ahead.

www.ingramcontent.com/pod-product-compliance
Ingram Content Group UK Ltd.
Pitfield, Milton Keynes, MK11 3LW, UK
UKHW020049250226
10904UKWH00039B/288

9 781923 545540